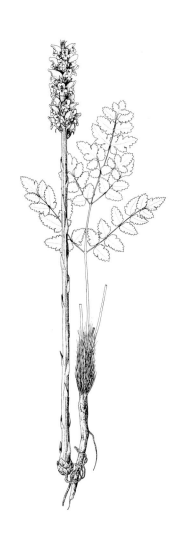

©**NATUURHISTORISCH** GENOOTSCHAP IN LIMBURG

CIP-GEGEVENS KONINKLIJKE BIBLIOTHEEK, DEN HAAG

Kreutz, C.A.J.

Orobanche : die Sommerwurzarten Europas : ein Bestimmungsbuch = the European broomrape species : a field guide / C.A.J. Kreutz ; mit Farbaufnahmen vom Autor / with colour photographs by the author. - Maastricht : Stichting Natuurpublicaties Limburg. - Ill., foto's
Bd. 1: Mittel- und Nordeuropa = Vol. 1: Central and Northern Europe.
Met index, lit. opg.
ISBN 90-74508-05-7 geb.
Trefw.: bremrapen ; Midden-Europa / bremrapen ; Noord-Europa.

AUSGABE / PUBLICATION: Stichting Natuurpublicaties Limburg
DISTRIBUTION / DISTRIBUTION: Publicatiebureau
Natuurhistorisch Genootschap in Limburg,
postbus 882, 6200 AW Maastricht, NL

Maastricht 1995

Niets uit deze uitgave mag worden verveelvoudigd en /of openbaar gemaakt door middel van druk, fotokopie, microfilm of op welke andere wijze dan ook, zonder voorafgaande schriftelijke toestemming van de uitgever.

Das Werk einschließlich aller seiner Teile ist urheberrechtlich geschützt. Jede Verwertung außerhalb der engen Grenzen des Urheberrechtsgesetzes ist ohne Zustimmung des Verlages unzulässig und strafbar. Das gilt insbesondere für Vervielfältigungen, Übersetzungen, Mikroverfilmungen und die Einspeicherung und Verarbeitung in elektronischen Systemen.

No part of this book may be reproduced in any form, by print, photoprint, microfilm or any other means, without written permission from the publishers.

Umschlag Vorderseite/frontcover
Farbaufnahme/colour photograph:
Orobanche alba mit Wirtspflanze/with host (*Thymus polytrichus*),
Tauplitz, Steiermark (A), 25-7-1989

Umschlag Rückseite/backcover
Farbaufnahmen/colour photographs:
Orobanche ceasia, Jois, Burgenland (A), 28-5-1994
Orobanche artemisiae-campestris, Bad Frankenhausen,
Kyffhäusergebirge (D), 16-6-1992
Verbreitungskarte/distribution map:
Orobanche alba

Illustration Seite 1/illustration page 1:
Orobanche alsatica mit Wirtzpflanze/with host (*Peucedanum cervaria*)

OROBANCHE
Die Sommerwurzarten Europas
The European broomrape species

Ein Bestimmungsbuch

A field guide

1

MITTEL- UND NORDEUROPA
CENTRAL AND NORTHERN EUROPE

C. A. J. Kreutz

mit Farbaufnahmen vom Autor / with colour photographs by the author

Orobanche ramosa, Gorazde, Bosnai Heceqovina (BA), 10-7-1987

Orobanche amethystea, Lozeron, Vercors (F), 29-5-1990

Orobanche picridis, Wijk aan Zee, Noord-Kennemerland (NL), 20-6-1989

Orobanche bartlingii, Ruprechtsstegen, Nördliche Frankenalb (D), 8-7-1994

Inhalt

6	1.	Vorwort
7		**ALLGEMEINER TEIL**
8	2.1	Einführung
8	2.2	Systematische Stellung
9	2.3	Taxonomie und Nomenklatur
11	2.4	Merkmale der Sommerwurzarten
14	2.5	Bestäubung
14	2.6	Keimung
14	2.7	Vermehrung
15	2.8	Morphologie
15	2.9	Wirtspflanzen
16	2.10	Standort (Lebensraum), Ökologie
16	2.11	Verbreitung
16	2.12	Gefährdung und Schutz
18	2.13	Adventive Vorkommen
18	2.14	Ernteschäden
19	2.15	Bekämpfung *Orobanche*-Arten in landwirtschaftlichen Kulturen
20	2.16	Literatur
21		**SPEZIELLER TEIL**
22	3.	Bestimmung der mittel- und nordeuropäischen *Orobanche*-Arten
22	3.1	Blütenbau
24	3.2	Übersicht von Fachausdrücken
24	3.3	Die Arten
25	3.4	Hinweise zur Bestimmung
25	3.5	Bestimmungsschlüssel der mittel- und nordeuropäischen *Orobanche*-Arten
32	4.	Beschreibung der Arten
32		SEKTION **TRIONYCHON**
32	4.1	*Orobanche arenaria*
36	4.2	*Orobanche caesia*
40	4.3	*Orobanche purpurea*
44	4.4	*Orobanche ramosa*
48		SEKTION **OROBANCHE**
48	4.5	*Orobanche alba*
52	4.6	*Orobanche alsatica*
56	4.7	*Orobanche alsatica* subsp. *mayeri*
60	4.8	*Orobanche amethystea*
64	4.9	*Orobanche artemisiae-campestris*
68	4.10	*Orobanche bartlingii*
72	4.11	*Orobanche caryophyllacea*
76	4.12	*Orobanche cernua*
80	4.13	*Orobanche coerulescens*
84	4.14	*Orobanche crenata*
88	4.15	*Orobanche cumana*
92	4.16	*Orobanche elatior*
96	4.17	*Orobanche flava*
100	4.18	*Orobanche gracilis*
104	4.19	*Orobanche hederae*
108	4.20	*Orobanche laserpitii-sileris*
112	4.21	*Orobanche lucorum*
116	4.22	*Orobanche lutea*
120	4.23	*Orobanche minor*
124	4.24	*Orobanche pallidiflora*
128	4.25	*Orobanche picridis*
132	4.26	*Orobanche rapum-genistae*
136	4.27	*Orobanche reticulata*
140	4.28	*Orobanche salviae*
144	4.29	*Orobanche teucrii*
148	4.30	*Orobanche variegata*
153	5.	Literatur
157	6.	Inhaltsverzeichnis
157	6.1	Verzeichnis der Sommerwurznamen
157	6.2	Index of broomrape names
157	6.3	Index van bremraapnamen
157	6.4	Index of scientific names
159	7.	Danksagung

Contents

6	1.	Preface
7		**GENERAL PART**
8	2.1	Introduction
8	2.2	Systematic position
9	2.3	Taxonomy and nomenclature
11	2.4	Characteristics of the broomrape species
14	2.5	Pollination
14	2.6	Germination
14	2.7	Propagation
15	2.8	Morphology
15	2.9	Host plants
16	2.10	Habitat, Ecology
16	2.11	Distribution
16	2.12	Threats and conservation
18	2.13	Adventive locations
18	2.14	Crop damage
19	2.15	Controlling *Orobanche* species in agricultural environments
20	2.16	Bibliography
21		**SPECIFIC PART**
22	3.	Identification of central and northern European *Orobanche* species
22	3.1	Structure of the flower
24	3.2	Glossary of terms
24	3.3	The species
25	3.4	Suggestions for identification
25	3.5	Key to *Orobanche* species in central and northern Europe
32	4.	Species descriptions
32		SECTION **TRIONYCHON**
32	4.1	*Orobanche arenaria*
36	4.2	*Orobanche caesia*
40	4.3	*Orobanche purpurea*
44	4.4	*Orobanche ramosa*
48		SECTION **OROBANCHE**
48	4.5	*Orobanche alba*
52	4.6	*Orobanche alsatica*
56	4.7	*Orobanche alsatica* subsp. *mayeri*
60	4.8	*Orobanche amethystea*
64	4.9	*Orobanche artemisiae-campestris*
68	4.10	*Orobanche bartlingii*
72	4.11	*Orobanche caryophyllacea*
76	4.12	*Orobanche cernua*
80	4.13	*Orobanche coerulescens*
84	4.14	*Orobanche crenata*
88	4.15	*Orobanche cumana*
92	4.16	*Orobanche elatior*
96	4.17	*Orobanche flava*
100	4.18	*Orobanche gracilis*
104	4.19	*Orobanche hederae*
108	4.20	*Orobanche laserpitii-sileris*
112	4.21	*Orobanche lucorum*
116	4.22	*Orobanche lutea*
120	4.23	*Orobanche minor*
124	4.24	*Orobanche pallidiflora*
128	4.25	*Orobanche picridis*
132	4.26	*Orobanche rapum-genistae*
136	4.27	*Orobanche reticulata*
140	4.28	*Orobanche salviae*
144	4.29	*Orobanche teucrii*
148	4.30	*Orobanche variegata*
153	5.	Bibliography
157	6.	Index
157	6.1	Verzeichnis der Sommerwurznamen
157	6.2	Index of broomrape names
157	6.3	Index van bremraapnamen
157	6.4	Index of scientific names
159	7.	Acknowledgements

Vorwort

Seit etwa 1973 beschäftige ich mich, übrigens nicht beruflich, mit europäischen Orchideen. In diesen zwanzig Jahren konnten fast alle Arten, Unterarten und Varietäten Europas an den natürlichen Standorten aufgefunden werden. Dabei habe ich diese Pflanzen studiert und fotografiert, und vor allem in den Niederlanden und in Deutschland wurden in Zusammenarbeit mit den Naturschutzbehörden viele Standorte geschützt. Über diese seltene und wunderschöne Pflanzenfamilie habe ich viele Artikel in niederländischen und deutschen botanischen Zeitschriften geschrieben, außerdem zwei Bücher über die Orchideen der Niederlande verfaßt. Zu diesem Zweck wurden alle Länder Europas und Vorderasiens (von Großbritannien bis zur Osttürkei und von Portugal bis Polen) besucht.

Viele *Orobanche*-Arten wachsen oft in den gleichen Biotopen wie Orchideen, wobei einige Sommerwurzarten in ziemlich schwer zugänglichen Gebieten vorkommen (zum Beispiel in Felsformationen). Sie sind interessant und selten und mit ihren farbenprächtigen Blüten genau so schön wie Orchideen. Außerdem gibt es, genau wie bei den Orchideen, problematische Arten. Etwa ab 1980 habe ich viele Orobanchen an ihren natürlichen Standorten studiert und fotografiert. Anfang der neunziger Jahre stellte sich heraus, daß ich fast alle europäischen Arten gefunden und fotografiert hatte, und so reifte der Plan, ein Buch über alle europäischen Arten herauszugeben. Dabei sollte jede Art mit guten Farbaufnahmen abgebildet werden. Ziemlich schwierig war es, noch fehlende Arten (wie zum Beispiel *Orobanche caesia* und *O. salviae*) zu finden, aber durch gezieltes Suchen konnten auch diese beiden Arten in Österreich und Süddeutschland aufgefunden werden. Auch war es ziemlich schwierig, alle Arten in ihrer Hauptblütezeit zu fotografieren, denn alle Sommerwurzarten weisen gewöhnlich eine (sehr) kurze Blütezeit auf.

In diesem Band sind erstmals alle mittel- und nordeuropäischen Arten beschrieben und mit Zeichnungen, Verbreitungskarten und Farbaufnahmen abgebildet. Ich hoffe, daß die Bestimmung der Sommerwurzarten jetzt leichter geworden ist und dadurch das Interesse an diesen wunderschönen und geheimnisvollen Pflanzen verstärkt wird.

Preface

Since 1973, European orchids have had my special, albeit non-professional interest. During this twenty year period I have been able to find most European species, sub-species and varieties in their natural habitat. I have studied and photographed these plants and - in close cooperation with environmental protection agencies in the Netherlands and in Germany - many locations have been declared protected areas as a result. I have published many articles on these rare and exquisite plants in Dutch and German botanic journals and written two books on the orchids of the Netherlands. This interest resulted in my travelling to most countries of Europe and the Near East (from the United Kingdom to eastern Turkey and from Portugal to Poland).

Many *Orobanche* species grow in the same habitats as orchids, although some grow in locations which are quite inaccessible, such as rock faces. Broomrapes are interesting, rare species and their richly coloured flowers are just as beautiful as those of orchids. Like orchid species, several are problematic. From 1980 on I have studied and photographed many *Orobanche* species in their natural habitat. In the early 1990s I found that I had located and photographed nearly all European species, so I planned to write a book on them. Each species was to be documented with quality colour photographs. Finding the missing species (like *Orobanche caesia* and *O. salviae*) was difficult, but a dedicated search resulted in their being located in Austria and southern Germany. Another complication was the need to photograph all species in their prime bloom, as most of them have a short flowering time.

This book documents for the first time all central and northern European species, using drawings, distribution maps and colour photographs. I hope it will make the identification of broomrape species much easier and that interest in this exquisite and mysterious group of plants will increase.

C.A.J. Kreutz
Landgraaf, November 1994

ALLGEMEINER TEIL GENERAL PART

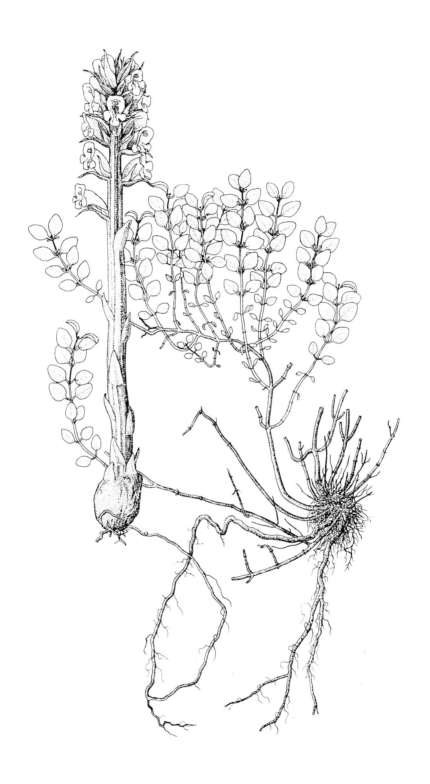

Orobanche alba mit Wirtspflanze (*Thymus pulegioides*)　　*Orobanche alba* with host (*Thymus pulegioides*)

2.1 EINFÜHRUNG

Die Sommerwurzgewächse (Orobanchen) haben die Botaniker schon immer fasziniert. Sie gehören mit den Orchideen zu den interessantesten Pflanzen der Welt; die meisten Arten sind mit ihren farbenprächtigen Blüten (von hellgelb oder violett bis zu leuchtendem Rot) genau so schön wie Orchideen. Bei den Orobanchen gibt es viele problematische Arten. Es werden zahlreiche Varietäten und Formen unterschieden (Gilli, 1966), die in den meisten Fällen ökologische Formen betreffen. Daß es so viele Varietäten und Formen gibt, hängt wahrscheinlich auch von der Art und Qualität der Wirtspflanze ab. Auch sind viele Arten sehr variabel in ihrer Erscheinungsform und Blütenfarbe. Bei einigen Arten sind Verfärbungen der gesamten Pflanze bekannt. Von fast allen Arten werden selten auch hellgelbe Exemplare gefunden, wodurch die Bestimmung manchmal schwierig wird. Die Blüten von Orobanchen, die an schattigen Stellen wachsen, sind meistens heller gefärbt (Blumenkrone und Narbe oft gelblich). Sommerwurzarten weisen gewöhnlich eine kurze Blütezeit auf. Meistens entwickeln sich aber über eine Zeit von etwa zwei bis drei Wochen immer neue Blütenstände, so daß die meisten Arten doch über eine längere Zeit beobachtet werden können. Fällt die Blütezeit in eine Trockenperiode, dann kommen meistens nur wenige Pflanzen zur Blüte und sind oft innerhalb einer Woche verblüht. Regnet es in dieser Periode, dann entwickeln sich meistens in sehr kurzer Zeit einige neue Blütenstände (Nachzügler). Der Wirt braucht nicht immer direkt neben dem Schmarotzer zu stehen, er kann auch einige Meter entfernt wachsen.

Durch all diese Faktoren sind die einzelnen Arten meistens schwer zu bestimmen und Fehler nicht auszuschließen! Daher haben diese Pflanzen weniger Botaniker interessiert, als zum Beispiel die Orchideen und Lilien. Im Gegensatz zu diesen beiden Familien gibt es über Orobanchen keine Arbeiten, die alle europäischen Arten mit verschiedenen Farbaufnahmen und Verbreitungskarten enthalten. Die meisten Beschreibungen, oft allerdings ohne fotografische Abbildungen, findet man in heimischen und internationalen Floren.

Das vorliegende Buch trägt dazu bei, alle Arten leicht und schnell zu bestimmen. Dies wird mit Hilfe eines Bestimmungsschlüssels, einer Zeichnung der Seitenansicht der Blüte, einer Zeichnung der Staubblätter (Staubfäden und Staubbeutel) und des Stempels (Griffel, Narbe und Fruchtknoten), sowie der Angabe des Verbreitungsgebietes (Areal) ermöglicht. Im Übrigen soll die Bestimmung durch fünf Farbaufnahmen (Habitus/Standort, Blütenstand, Ausschnitt vom Blütenstand, Blüte von der Seite und Voderansicht einer Einzelblüte) unterstützt werden.

Dieser Band enthält alle nord- und mitteleuropäischen Arten. Die Arten Südeuropas werden im zweiten Band behandelt. Mit dem Erscheinen des zweiten Bandes wird erstmals ein Übersicht aller europäischen *Orobanche*-Arten vorliegen.

2.2 SYSTEMATISCHE STELLUNG

Die Familie der *Orobanchaceae* umfaßt weltweit 16 Gattungen mit über 200 Arten. Nach Chater & Webb in der *Flora Europaea* (1972) kommen in Europa vier Gattungen vor: *Boschniakia* (*B. rossica*), *Cistanche* (*C. phelypaea* & *C. salsa*), *Orobanche* und *Phelypaea* (*P. coccinea* & *P. boissieri*).

Cistanche phelypaea wächst hauptsächlich im Mittelmeerraum und schmarotzt auf holzigen Gänsefußgewächsen. So sind mehrere Standorte an der Küste von Portugal und Spanien bekannt, wo sie an feuchten Stellen in den Sanddünen wächst. Sie tritt seltener im mittleren mediterranen Bereich auf, aber sie ist an vielen Stellen im östlichen Mittelmeerraum, zum Beispiel am Toten Meer in Israel zu finden. *Boschniakia rossica* hat ihr Verbreitungsgebiet hauptsächlich in Asien. Die Arten der Gattung *Phelypaea* sind vor allem in Rußland (Krim und Kaukasus), im ehemaligen Süd-Jugoslawien und in Anatolien verbreitet.

2.1 INTRODUCTION

The broomrapes (*Orobanche*) have always fascinated botanists. With the orchids, broomrapes are among the most interesting plants in the world, and most species, with their colourful flowers (from bright yellow to violet or bright red), are just as beautiful as orchids. But the broomrapes include many problematic species. Many varieties and forms have been distinguished (Gilli, 1966), which in many cases are ecological forms. The fact that so many varieties and forms exist is probably linked to the types and qualities of the host plants as well. Many species also show a wide variety in their appearance and the colour of their flowers. In some species, the colour of the whole plant is known to be variable. Bright yellow forms are occasionally found of almost all species, which may make identification difficult. Flowers of species growing in shady places are usually of a lighter colour (corolla and stigma frequently yellowish). Broomrapes usually have a short flowering time. Normally, however, new inflorescences continue to be produced over a period of two to three weeks, so that most species can still be observed over a reasonable period. If the flowering time coincides with a time of drought, few plants may develop and these will wilt within a week. Rain during this period usually causes a rapid growth of new plants (laggards). The host plant does not necessarily have to be close to the parasite; it can also grow several metres away.

Because of all these factors, the different species are not easy to identify, and errors cannot be excluded! Because of the problems of identification, the plants used to be of less interest to botanists than orchids and lilies. In contrast to both these families, no books exist which incorporate all European *Orobanche* species and include full colour photographs and distribution maps. Most descriptions, without photographs, have been published in national and international floras.

The present book is intended as a contribution to the easy and rapid identification of all species. Identification is supported by a key, by drawings of side views of the corolla, as well as of the stamens (filaments and anthers) and the carpel (style, stigma and ovary), and by specification of the distribution (range). Five colour photographs (appearance/habitat, inflorescence, detail of inflorescence, side and front views of an individual flower) further facilitate identification.

All northern and central European species have been incorporated in this book. The southern European (especially Mediterranean) species will be incorporated in part two. Both volumes together will constitute the first European review on all species of *Orobanche*.

2.2 SYSTEMATIC POSITION

The family of *Orobanchaceae* comprises 16 genera with more than 200 species worldwide. According to Chater & Webb in the *Flora Europaea* (1972), 4 genera are represented in Europe: *Boschniakia* (*B. rossica*), *Cistanche* (*C. phelypaea* & *C. salsa*), *Orobanche* and *Phelypaea* (*P. coccinea* & *P. boissieri*).

Cistanche phelypaea grows mainly in the Mediterranean area and parasitizes woody goosefoot species. There are several known locations along the Portuguese and Spanish coast, where they grow on moist places in the sand dunes. They are rarely seen in the central Mediterranean area, but appear again in the eastern Mediterranean region, for instance on the shore of the Dead Sea in Israel. *Boschniakia rossica* grows mainly in Asia. The species of the genus *Phelypaea* grow mainly in Russia (Krim and Caucasus), in the southern part of former Yugoslavia and Anatolia.

Cistanche phelypaea, Coutinho (L.), Arvot Yeriho (IL), 20-3-1992

Orobanche foetida Poiret, Vila do Bispo, Algarve (P), 12-4-1994

Die Gattung *Orobanche* umfaßt über 150 Arten, wovon die meisten in den Mittelmeerländern vorkommen. Sie steht morphologisch und taxonomisch den *Scrophulariaceae* und *Gesneriaceae* nahe (Pusch & Barthel, 1992). Im Gegensatz zu den *Scrophulariaceae* und den *Gesneriaceae* besitzen die *Orobanchaceae* aber immer vollkommen oberständige Fruchtknoten, beziehungsweise einen stets einfächerigen Fruchtknoten.

Nach Gilli (1966) gehören den nord- und mitteleuropäischen *Orobanche*-Arten zwei Sektionen an:

1. *Trionychon* Wallr. Blüte mit zwei seitlichen kleinen, gegenständigen Vorblättern. Kelch röhrig, ungeteilt, 4-, seltener 5-zähnig. Blütenkrone blau oder violett (*Orobanche arenaria*, *O. caesia*, *O. purpurea* und *O. ramosa*). Diese Arten sind alle ziemlich klein, durchschnittlich etwa 20 Zentimeter hoch und sie haben alle blaugefärbte Blüten. *O. ramosa* ist die einzige mitteleuropäische Art, die mehrere Blütenstände entwickeln kann.

Für Arten dieser Sektion wird manchmal der Gattungsname *Phelipanche* Pomel (unter anderem Coste, 1937; Dostal, 1989) verwendet (*Phelipanche arenaria*, *P. caesia*, *P. purpurea* und *P. ramosa*).

2. *Orobanche* L. (= *Osproleon* Wallr.) Stengel immer einfach. Blüte ohne Vorblätter. Kelch in zwei seitliche, ein- oder zweizähnige, oft vorn (selten auch hinten) im unteren Teil miteinander verbundene Hälften gespalten. Blütenkrone meistens gelb, braun oder rot gefärbt. Dies betrifft alle übrigen Arten.

2.3 TAXONOMIE UND NOMENKLATUR

Bei einigen Arten besteht noch keine Übereinstimmung über die Taxonomie und Nomenklatur. Einige Autoren fassen Arten zusammen, andere Autoren haben Unterarten in den Artrang erhoben. Von vielen Arten wurden verschiedene Varietäten und Formen beschrie-

The *Orobanche* genus comprises more than 150 species, most of which grow in the Mediterranean region. *Orobanche* species are morphologically and taxonomically close to the *Scrophulariaceae* and *Gesneriaceae* (Pusch & Barthel, 1992). Unlike *Scrophulariaceae* and *Gesneriaceae*, the *Orobanchaceae* always have completely superior ovaries and single capsules.

According to Gilli (1966) the *Orobanche* species of northern and central Europe belong to two sections:

1. *Trionychon* Wallr. Corolla with two small, lateral, opposite bracteoles. Calyx tubular, entire, 4-, rarely 5-dentate. Corolla white, blue or violet (*Orobanche arenaria*, *O. caesia*, *O. purpurea* and *O. ramosa*). These species are all relatively small, approximately 20 cm tall, and all have blue flowers. *O. ramosa* is the only central European species developing multiple inflorescences.

Species of this section are sometimes given the genus name *Phelipanche* Pomel (*Phelipanche arenaria*, *P. caesia*, *P. purpurea* and *P. ramosa*), e.g. by Coste, 1937; Dostal, 1989.

2. *Orobanche* L. (= *Osproleon* Wallr.). Stem always simple. Flower without bracteoles. Calyx divided into two lateral halves, 1- or 2-dentate, often fused below at the front (rarely at the back). Corolla usually yellow, brown or red. This applies to all remaining species.

2.3 TAXONOMY AND NOMENCLATURE

For some species there is still no agreement on the taxonomy and nomenclature. Some authors have combined species, others have elevated subspecies to species level. Various varieties and forms have been described of many species (see Beck von

ben (vergleiche Beck von Mannagetta, 1890; Beck-Mannagetta 1930; Gilli, 1966). Im vorliegenden Buch werden die reinen Arten und Unterarten besprochen! Wichtige Varietäten und Formen werden kurz bei der Nominatform beschrieben.

Mit Ausnahme von *O. bartlingii, O. pallidiflora, O. cumana* und *O. alsatica* subsp. *mayeri* subsp. nov. wird in diesem Buch die Taxonomie und Nomenklatur der Arten weitgehend nach Gilli (1966) übernommen. *Orobanche reticulata* subsp. *reticulata* und *O. reticulata* subsp. *pallidiflora* weisen eine große Zahl verschiedener Merkmale auf, sowohl morphologisch als geografisch. Genau wie in den meisten osteuropäischen Publikationen (unter anderem Novopokrovskij *et al.*, 1950; Jaudzeme, 1959; Eilart *et al.*, 1973; Jankeviciene, 1976; Fedorov, 1981) ist im vorliegenden Buch *O. reticulata* subsp. *pallidiflora* als eigenständige Art aufgefüht (*O. pallidiflora* Wimmer et Grabowski 1829).

Orobanche bartlingii wird hier als Art aufgenommen. Bei einem Vergleich mit *O. alsatica* stellt sich heraus, daß *O. bartlingii* eine zierliche Pflanze mit meist kleineren Blüten ist, die Krümmung des Blütenrückens nicht gleichmäßig und im mittleren Teil durch eine abgeflachte Stelle unterbrochen ist, sich die Ansatzstelle der Staubblätter näher am Grund der Kronröhre befindet, der Griffel kahl oder selten mit einzelnen Drüsenhaaren besetzt ist, und schließlich daß sie eine frühere Blütezeit hat und auf einem anderen Wirt schmarotzt. In allen botanischen Werken jüngster Zeit wird *O. bartlingii* daher als eigenständige Art aufgefaßt.

Orobanche cumana wird hier von *O. cernua* abgetrennt und als eigene Art aufgenommen, weil sie fast immer auf Sonnenblumen, seltener auf Tabak (*Nicotiana tabacum*) und Tomaten (*Solanum lycopersicum*) parasitiert. *O. cernua* wächst dagegen fast ausschließlich in natürlichen Vegetationseinheiten und schmarotzt dort vor allem auf *Artemisia campestris*. Die Erhebung dieser Sippe in den Artrang wird von vielen Autoren und Sommerwurzspezialisten (Novopokrovskij *et al.*, 1950; Madalski, 1967; Savulescu *et al.*, 1963, Ter Borg *et al.*, 1986; Wegmann *et al.*, 1991) angeregt. *O. cumana* unterscheidet sich unter anderem von *O. cernua* durch ihren kräftigen Habitus und ihren zierlichen und lockeren Blütenstand. Außerdem sind die Blüten von *O. cumana* meistens größer als 15 mm (die von *O. cernua* sind meistens kleiner als 15 mm).

Orobanche alsatica subsp. *mayeri*, siehe Seite 56 (*O. alsatica* var. *mayeri* Suessenguth et Ronniger 1942; *O. mayeri* (Suessenguth et Ronniger) K. & F. Bertsch 1948), die auf der Schwäbischen Alb in Baden-Württemberg und in Bayern (Maingebiet am Kalbenstein bei Karlstadt) in Deutschland vorkommt, wird in diesem Buch wegen der zu geringen Unterschiede gegenüber *O. alsatica* und *O. bartlingii* nicht als eigenständige Art bewertet aber nur als eine Unterart betrachtet. Weil diese Sippe in einigen deutschen Exkursionsfloren jedoch noch immer als Art aufgefaßt wird, ist sie in diesem Buch ausführlich beschrieben worden. Sie unterscheidet sich von *O. alsatica* durch ihre meist reingelben Stengel und Blumenkronen, ihre meistens vorne nicht verwachsenen Kelchhälften, ihre etwas kleineren Blüten (12-15 mm) und ihre Wirtspflanze (*Laserpitium latifolium*).

Von *Orobanche purpurea* ist die Varietät *bohemica* (Celak.) Beck 1890 beschrieben, durch manche Autoren als eigene Art (*O. bohemica* Celakakovski 1874.; *Phelipaea bohemica* (Celak.) Holub) abgetrennt. *O. purpurea* var. *bohemica* unterscheidet sich von *O. purpurea* durch folgende Merkmale: Die Pflanze ist sehr kräftig (bis zur 70 cm hoch) und in allen Teilen meist stärker gefärbt (dunkelviolett) als *O. purpurea*; ihr Blütenstand ist ziemlich lang, viel- und dichtblütig. Die Blütenkrone ist 20 bis 25 mm lang. Der Stengel ist reichlicher beschuppt als bei *O. purpurea* und die Staubbeutel sind kahl. Sie schmarotzt nur auf *Artemisia campestris*. Ihr Verbreitungsgebiet liegt vor allem im östlichen Teil von Mitteleuropa (unter anderem in Niederösterreich und der Tschechischen Republik), sie kommt aber auch zerstreut in Ostdeutschland, Norditalien (Tirol) und in der Schweiz vor. *O. purpurea* var. *bohemica* wurde von Velká hora in der Nähe von Karlstejn (25 km SW Prag) beschrieben. Zazvorka (Pruhonice, Tschechische Republik) hat diesen Standort ab 1975 jedes Jahr besucht, fand aber *O. purpurea* var. *bohemica* nur in den Jahren 1981 und 1985. In vorliegendem Buch

Mannagetta, 1890; Beck-Mannagetta 1930; Gilli, 1966). The present book discusses the pure species and subspecies. Important varieties and forms are briefly described together with the typical form.

With the exception of *O. bartlingii, O. pallidiflora, O. cumana,* and *O. alsatica* subsp. *mayeri* subsp. nov. the present book largely uses the taxonomy and nomenclature according to Gilli (1966). *Orobanche reticulata* subsp. *reticulata* and *O. reticulata* subsp. *pallidiflora* show a large number of different characteristics, both in terms of morphology and geography. As in most eastern European publications (e.g. Novopokrovskij *et al.*, 1950; Jaudzeme, 1959; Eilart *et al.*, 1973; Jankeviciene, 1976; Fedorov, 1981) the present book lists *O. reticulata* subsp. *pallidiflora* as a separate species (*O. pallidiflora* Wimmer et Grabowski 1829).

Orobanche bartlingii is listed as a separate species. In comparison to *O. alsatica*, *O. bartlingii* is a more delicate plant with smaller flowers; the dorsal line of the corolla is not continuously curved, but is interrupted by a straight section in the middle; the stamens are inserted closer to the base of the corolla-tube; the style is glabrous or rarely covered with single glandular hairs; it flowers earlier and is parasitic on a different host. All recent botanical works list *O. bartlingii* as a separate species as well.

Orobanche cumana has been separated from *O. cernua* and treated as a species here, because it is mostly parasitic on sunflower cultures, rarely on tobacco (*Nicotiana tabacum*) and tomato (*Solanum lycopersicum*). *O. cernua* grows almost exclusively in natural vegetation and is parasitic especially on *Artemisia campestris*. Classification of this variety as a species has been suggested by many authors and broomrape specialists (Novopokrovskij *et al.*, 1950; Madalski, 1967; Savulescu *et al.*, 1963; Ter Borg *et al.*, 1986; Wegmann *et al.*, 1991). *O. cumana* differs from *O. cernua* in its stout build. The plants are more graceful and have a lax inflorescence. The flowers are longer than 15 mm (while those of *O. cernua* are usually shorter than 15 mm).

Orobanche alsatica subsp. *mayeri*, see page 56 (*O. alsatica* var. *mayeri* Suessenguth et Ronniger 1942; *O. mayeri* (Suessenguth et Ronniger) K. & F. Bertsch 1948), which grows on the Schwäbische Alb in Baden-Württemberg and in the Main area (Kalbenstein) near Karlstadt in Bavaria in Germany, is not given the status of a separate species in the present book but is classified as a subspecies as it differs too little from *O. alsatica* and *O. bartlingii*. It has been included here nevertheless, because it is still listed as a separate species in several German field guides. It differs from *O. alsatica* in its mostly pure yellow stem and corolla, its free calyx segments, its slightly smaller flowers (12-15 mm) and its host (*Laserpitium latifolium*).

Of *Orobanche purpurea* the variety *bohemica* (Celak.) Beck 1890 has been defined by some authors (*O. bohemica* Celakakovski 1874.; *Phelipaea bohemica* (Celak.) Holub) as a separate species. *O. purpurea* var. *bohemica* differs from *O. purpurea* in the following characteristics. The plant is very stout (up to 70 cm tall) and is more intensely coloured than *O. purpurea* in all parts (dark violet); its inflorescence is long and dense, with numerous flowers. The stem is more densely scaled than that of *O. purpurea* and the anthers are glabrous. It is parasitic only on *Artemisia campestris*. Its range is mainly in the eastern part of central Europe (Austria (Niederösterreich) and the Czech Republic), but it also occurs sporadically in eastern Germany, northern Italy (Alto Adige) and in Switzerland. *O. purpurea* var. *bohemica* has been described from Velká hora near Karlstejn (25 km south-west of Prague). Zazvorka (Pruhonice, Czech Republic) has visited this location annually from 1975 onwards, but found *O. purpurea* var. *bohemica* only in 1981 and 1985. The present book does not classify this variety as a separate species. Zazvorka, who has studied the variety at sev-

wird diese Sippe nicht als eine eigenständige Art bewertet. Diese Meinung vertritt auch Zazvorka, der die Varietät an einige Stellen in der Tschechischen Republik studiert hat (schrift. Mitteilung, 1994).
Von *Orobanche minor* sind viele Varietäten bekannt; die wichtigsten sind hauptsächlich in West-Europa verbreitet.
Von Südengland (unter anderem Südwales, Kent, Dorset, Devon, Cornwall), den Kanal-Inseln und Westfrankreich ist *Orobanche minor* var. *maritima* (Pugsley) Rumsey & Jury 1989 (*O. maritima* Pugsley 1940; *O. amethystea* auct. non Thuill.) beschrieben (2n=38). Pflanzen dieser Varietät wurden erstmals durch Hore 1845 in den Dünen bei Whitsand Bay (Cornwall) gefunden und als *O. amethystea* bestimmt. Bei dieser Varietät ist der Mittellappen der Unterlippe der Blütenkrone nierenförmig und größer als die beiden Seitenlappen. Die Blüten sind etwa 6 bis 10 mm groß, die untersten Blüten sind meistens gestielt. Die zwei halbkugeligen Lappen der Narbe sind teilweise miteinander verbunden. Diese Sippe schmarotzt auf *Daucus carota*, selten auf *Plantago coronopus* und *Ononis repens*.
Orobanche minor var. *compositarum* Pugsley 1940 hat kleinere Blüten als *O. minor* (Blütenkrone 12 bis 18 mm lang und 3,5 bis 5 mm breit), die aufrechter stehen, bleicher violett gefärbt und spärlicher mit Drüsenhaaren besetzt sind; die Staubblätter an der Basis sind dichter behaart. Diese Varietät parasitiert auf *Asteraceae*-Arten, hauptsächlich auf *Crepis virens, Hypochoeris radicata, Tripleurospermum inodorum, Carduus nutans* und *Senecio greyii* und wächst vor allem in Ost-Anglia in England. Jones (1989) kommt auf Grund seiner Untersuchungen zu dem Ergebnis, daß die beiden Varietäten '*maritima*' und '*compositarum*' sich nicht von *O. minor* unterscheiden; nur bei *O. minor* var. *maritima* ist der Habitus meistens kräftiger und die Blütenkrone etwas länger.
Die für Südengland angegebene *Orobanche amethystea* (Philp, 1982) ist *O. minor* zuzuordnen. Einige Populationen, die in Kent wachsen, kommen in einigen Merkmalen *O. amethystea* zwar nahe, aber *O. amethystea* hat größere Blüten, wobei die Blütenkrone eine auffallende zweilappige, aufrecht-abstehende Oberlippe aufweist und die Staubblätter höher über dem Grund der Kronröhre eingefügt sind. Außerdem wachsen diese Pflanzen auf vielen verschiedenen Wirtspflanzen (Rumsey & Jury, 1991).
Von Sommerwurzarten sind bisher keine Hybriden bekannt. Es bestehen zwar einige diesbezügliche Angaben, die aber der Überprüfung bedürfen.
Die im vorliegenden Buch verwendeten lateinischen Namen der Wirts- und Begleitpflanzen richten sich im allgemeinen nach der *Liste der Gefäßpflanzen Mitteleuropas* (Ehrendorfer, 1973).

2.4 MERKMALE DER SOMMERWURZARTEN

Orobanchen sind chlorophyllfreie, ein- bis mehrjährige, auf den Wurzeln anderer Pflanzen lebende Vollschmarotzer, die spargelartig aus dem Boden hervortreten. Die Wurzeln der Orobanchen sind teils als Saugwurzeln, teils als Adventivwurzeln ohne Wurzelhaare ausgebildet. Die aufrechten, einfachen Stengel sind an ihrem Grund verdickt und von gelblicher, brauner oder violetter Farbe. Blattgrün fehlt vollständig, wodurch sie nicht in der Lage sind, durch Photosynthese organische Substanzen zu erzeugen. Die reduzierten Blätter sind nur in Form von Schuppen vorhanden. Die Blüten stehen in endständigen Ähren in den Achseln schuppenförmiger Tragblätter. Der Kelch ist röhrig, besteht aus zwei seitlichen ein- bis zweizähnigen, vorne oft verwachsenen Teilen. Die Blumenkrone ist ebenfalls röhrig und weist eine zweispaltige oder ungeteilte Oberlippe sowie eine dreilappige Unterlippe auf. Die vier Staubblätter sind verschieden hoch eingefügt. Der Fruchtknoten ist einfächerig, die unterschiedlich gefärbte Narbe läßt zwei kugelige oder ovale Lappen erkennen.
Die Arten der Sektion *Trionychon* sind diploid (2n = 24) (Weber, 1976). Mit Ausnahme von *Orobanche gracilis* sind alle in Mitteleuropa vorkommenden Arten der Sektion *Orobanche* mit n = 19 resp. 2n = 38 diploid. *O. gracilis* ist tetra- und hexaploid, außerdem auch noch aneusomatisch (2n = 73-91, 112-116).

eral locations in the Czech Republic, agrees with this point of view (personal communication, 1994).
Many varieties of *Orobanche minor* are known; the most important ones grow mainly in western Europe.
Orobanche minor var. *maritima* (Pugsley) Rumsey & Jury 1989 (*O. maritima* Pugsley 1940; *O. amethystea* auct. non Thuill.) has been described from southern England (e.g. South Wales, Kent, Dorset, Devon, Cornwall), the Channel Islands and western France (2n=38). Plants of this variety were found for the first time by Hore in 1845 in the dunes near Whitsand Bay (Cornwall) and were identified as *O. amethystea*. The middle lobe of the lower lip of the corolla is reniform and larger than the side lobes. The flowers are approximately 6-10 mm, the lower ones usually pediculate. The two hemispherical lobes of the stigma are partly fused. It parasitizes on *Daucus carota*, rarely on *Plantago coronopus* and *Ononis repens*.
Orobanche minor var. *compositarum* Pugsley 1940 has smaller flowers than *O. minor* (corolla 12-18 mm long and 3.5-5 mm wide), which are more erect, of a paler violet and more sparsely glandular; stamens near the base are more densely pubescent. This variety parasitizes on *Asteraceae* species in particular on *Crepis virens, Hypochoeris radicata, Tripleurospermum inodorum, Carduus nutans* and *Senecio greyii*, and grows mainly in East-Anglia (England). Based on his investigations, Jones (1989) concludes that neither of the varieties '*maritima*' and '*compositarum*' can be distinguished from *O. minor*, the only difference being that *O. minor* var. *maritima* usually has a stouter build and a somewhat longer corolla.
The specimens of *Orobanche amethystea* (Philp, 1982) described for southern England should be classified as *O. minor*. The characteristics of a few populations growing in Kent come close to *O. amethystea*, but *O. amethystea* has larger flowers and its corolla has a patent, 2-lobed, spreading upper lip, with the stamens inserted higher above the base of the corolla-tube. Furthermore, these plants are parasitic on many different hosts (Rumsey & Jury, 1991).
So far, no hybrids of *Orobanche* are known. Several claims have been made, but these lack confirmation.
All Latin names of the host and other plants in this book are in accordance with the *Liste der Gefäßpflanzen Mitteleuropas* (Ehrendorfer, 1973).

2.4 CHARACTERISTICS OF THE BROOMRAPE SPECIES

Orobanche species are annual or perennial plants without chlorophyll, parasitic on roots of other plants, and emerging from the soil in an asparagus-like fashion. The roots of broomrapes are partly suckers, partly adventitious roots without root hairs. Stems are erect, single, thicker below and yellowish, brown or violet. Chlorophyll is completely absent, resulting in a total inability to produce organic substances through photosynthesis. Leaves are reduced to scale leaves. The flowers grow in single spikes from the axils of scale-like bracts. The calyx is tubular, consisting of two opposite 1- or 2-dentate parts, frequently fused at the front. The corolla is also tubular and has a bifid or entire upper lip, and a three-lobed lower lip. The four stamens are inserted at varying levels above the base of the corolla-tube. The ovary is 1-celled; the variably coloured stigma shows two spherical or oval lobes.
The species of the section *Trionychon* are diploid (2n=24) (Weber, 1976).
All central European species of the section *Orobanche* are diploid (n=19 or 2n=38), except *Orobanche gracilis*. *O. gracilis* is tetraploid or hexaploid and aneusomatic as well (2n=73-91, 112-116).

Orobanche alsatica subsp. *mayeri* (Bestäubung / pollination), Hechingen, Schwäbische Alb (D), 6-7-1994

Orobanche flava, Lunz am See, Ybbstaler Alpen (A), 9-7-1994

Orobanche minor mit einem Teil der Wirtspflanze, hier (and part of its host plant, in this case) *Trifolium pratense*, Landgraaf, Zuid-Limburg (NL), 18-7-1993

Orobanche salviae auf / on *Salvia glutinosa*, Markschellendorf, (D), 29-7-1993
Wurzelpol des Schmarotzers dringt in den Wirt ein (radical tuber infiltrating host's roots)

Orobanche arenaria (Stempel und Staubblätter / carpel and stamens), Schloßböcklheim, Hunsrück (D), 1-7-1989

Orobanche gracilis (Stempel und Staubblätter / carpel and stamens), Corvara in Badia, Dolomiti (I), 6-7-1986

Orobanche sanguinea C. Presl, Portintho, Sardegna (I), 26-4-1990

Orobanche densiflora Salzm., Aljezur, Algarve (P), 14-4-1994

2.5 BESTÄUBUNG

Die Sommerwurzarten sondern am Grunde der Staubblätter an lebhaft gefärbten Stellen Nektar ab. Die meisten Arten werden von Hummeln und Bienen bestäubt. Honigraub durch Anbeißen der Blumenkrone am Grund wurde mehrfach beobachtet. Bei einigen Arten tritt Selbstbestäubung auf. Hierbei wachsen die Staubblätter, bis die Antheren (Staubbeutel) die Narbe berühren (*Orobanche elatior* und *O. purpurea*) oder sich der obere Teil des Griffels mit der Narbe zu den Antheren hinabkrümmt (*O. ramosa*).

Als Lockmittel für Insekten dient vielen *Orobanche*-Arten ein von der Blumenkrone ausströmender Geruch (Gilli, 1966). Bei einigen Arten ist dieser Geruch angenehm, wie zum Beispiel bei *O. caryophyllacea* und *O. gracilis*, die beide nach Nelken duften, oder unangenehm (*O. rapum-genistae*). Jones (1991) studierte die Bestäuber vieler britischer *Orobanche*-Arten.

2.6 KEIMUNG

Die Keimung erfolgt nur, wenn die Samen der Sommerwurz Kontakt zu den Wurzeln der Wirtspflanze erhalten, genauer sie wird durch gewisse chemische Reizstoffe aus den Wurzeln des Wirtes ausgelöst. Der Keimling setzt sich mit seinem fadenförmigen Wurzelgebilde an der Wurzeloberfläche des Wirtes fest und dringt in die Wurzel ein, bis der Anschluß an die Gefäßbahnen erreicht ist. Es entsteht eine unterirdische Knolle, wobei Parasit und Wirt direkt miteinander verbunden sind. An dieser Knolle entstehen die Schuppenblätter und der Blütenstand. Dem Wirt werden Wasser, Mineralien (Baustoffe) und organische Nährstoffe entzogen. Die Wirtspflanzen bleiben in ihrer Entwicklung zurück und kommen meistens nicht zur Blüte; sie werden aber selten gänzlich zugrunde gerichtet. Die meisten Arten benötigen von der Aussaat bis zur Blüte einige Monate bis mehrere Jahre. Wenn die Schmarotzer an die Oberfläche gelangen, können sie innerhalb einiger Wochen blühen. *Orobanche minor* und *O. crenata* bilden schon im ersten Jahr, nachdem die Samen sich verbreitet haben, blühende Pflanzen.

Wenn der Parasit zu wenig Nahrung vom Wirt erhält, kommt er nicht zur Blüte und stirbt frühzeitig ab. Am Ende der Blütezeit ist der Kontakt zwischen der *Orobanche* und dem Wirt meistens unterbrochen und die Sommerwurz zehrt von eigenen Reserven. Nach der Blütezeit verwelkt der Blütenstand schnell. Manchmal bleibt in den Wurzeln des Wirtes Gewebe des Schmarotzers erhalten, und daraus kann sich im nächsten Jahr ein neuer Blütenstengel entwickeln. Der Parasit ist dann ausdauernd. Über die Lebensdauer vieler Arten ist aber nur wenig bekannt.

Einige *Orobanche*-Arten sind leicht zu züchten (zum Beispiel *Orobanche cumana*, *O. hederae*, *O. minor*, *O. picridis*, *O. crenata* und *O. ramosa*). Bei anderen Arten ist es viel schwieriger.

2.7 VERMEHRUNG

Die meisten Arten tragen rund 20 Blüten und produzieren in jeder Fruchtkapsel etwa 500-5000 Samen. Große Pflanzen, wie zum Beispiel *Orobanche crenata* oder *O. reticulata*, können pro Pflanze mehrere hunderttausend Samen erzeugen.

Bei Orobanchen, die auf Kulturpflanzen parasitieren, könnten theoretisch in dem darauffolgenden Jahr unter günstigen Verhältnissen etwa eine Million Pflanzen zur Blüte kommen, wenn man von hundert blühenden Pflanzen ausgeht, bei denen sich etwa jeder zehnte Samen entwickelt. Aber in der Natur trifft dies nicht zu. Die meisten Arten sind selten und von den vielen Samen erreichen nur sehr wenige die Wurzeln der geeigneten Wirte (auch Orchideen produzieren in ihren Samenkapseln eine ungeheure Menge von Samen und bei dieser

2.5 POLLINATION

The broomrape species secrete nectar from coloured spots at the base of the stamens. Most species are pollinated by bumblebees and bees. Insects have been observed to steal nectar by biting open the base of the corolla. Some species are self-pollinating. This is made possible by the stamens continuing to grow until their anthers come to touch the stigma (*Orobanche elatior* and *O. purpurea*) or by the upper part of the style with the stigma bending towards the anthers (*O. ramosa*).

To lure insects, most *Orobanche* species use a scent emitted by the corolla (Gilli, 1966). Some species have a pleasant scent, like *O. caryophyllacea* and *O. gracilis*, which smell like carnations, some emit an unpleasant smell (*O. rapum-genistae*). Jones (1991) has studied the pollinators of many British *Orobanche* species.

2.6 GERMINATION

Germination will take place only when the seed of the broomrape makes contact with the roots of the host plant, stricktly speaking germination is triggered by certain chemical signals released by the roots of the host plant. The seed attaches itself to the root surface of the host, using its own filiform root, and penetrates until it reaches the host's vascular system. An underground tuber develops, which constitutes a direct link between parasite and host. It is from this tuber that the scale leaves and the inflorescence develop. Water, minerals and organic nutrients are extracted from the host. The development of the host is stunted and it rarely flowers; total destruction of the host is rare, however. Most *Orobanche* species require between a few months and several years from germination to flowering. When the parasites rise above the soil surface, they can flower within a few weeks. *Orobanche minor* and *O. crenata* develop flowering plants in the first year after the seeds have been disseminated.

If the parasite fails to obtain sufficient nutrients from the host, it will not flower and will wither prematurely. At the end of the flowering time, the contact between *Orobanche* and its host is usually interrupted; the broomrape now lives on its own reserves. After the flowering period the inflorescence withers quickly. Sometimes tissue of the parasite remains lodged in the roots of the host and a new inflorescence can develop the next year. The parasite is then perennial. Little is known about the life span of many species.

Some *Orobanche* species are easy to cultivate (for example *Orobanche cumana*, *O. hederae*, *O. minor*, *O. picridis*, *O. crenata* and *O. ramosa*). Other species are much more difficult in this respect.

2.7 PROPAGATION

Most species have approximately 20 flowers and produce approximately 500-5.000 seeds in each capsule. Large plants, for instance *Orobanche crenata* or *O. reticulata*, can produce several hundred thousand seeds per plant.

One hundred *Orobanche* plants growing on cultivated hosts might, under favourable circumstances, produce approximately one million flowering plants in the next year, if one in ten seeds were to develop. But this does not happen in nature. Most species are rare and of the many seeds only very few reach the roots of a suitable host (Orchids too produce enormous numbers of seeds in their capsules and their development after the first germination is only possible when the

Pflanzenfamilie ist eine Weiterentwicklung nach der ersten Keimphase nur möglich, wenn der Samen im Boden vom richtigen Wurzelpilz infiziert wird). Die Samen der Orobanchen sind sehr klein und leicht (ca. 0,001 mg schwer). Dieses Samengewicht entspricht ungefähr dem der leichtesten Orchideensamen (Gilli, 1966). Sie dringen deshalb leicht in tiefere Erdschichten ein. Die Verbreitung findet durch den Wind statt. Im Endosperm ist Öl enthalten, so daß die Keimfähigkeit lange erhalten bleibt und bei manchen Arten etwa 10 bis 12 Jahre betragen kann.

2.8 MORPHOLOGIE

Die meisten Orobanchen besitzen einen einzigen Blütenstand. Bei wenigen anderen Arten, wie zum Beispiel *Orobanche ramosa*, die auch in Mitteleuropa vorkommt, ist der Stengel meistens verzweigt und hat mehrere Blütenstände. Meistens wachsen mehrere Exemplare gemeinsam; so wurde von *O. variegata* eine Gruppe von 78 Pflanzen beobachtet. Der Blütenstand ist meistens dichtblütig und ährenförmig, bei manchen Arten, wie zum Beispiel bei *O. purpurea* und *O. hederae* lockerblütig, wobei etwa zwei Drittel des gesamten Stengels mit Blüten besetzt ist.

Wenn man Orobanchen im Gelände findet, ist es wichtig, die Farbe der Narbe und Blüten (Krone) und vor allem den Wirt zu bestimmen. Dennoch können Narbe und Blüte manchmal eine von der Nominatform abweichende Farbe aufweisen. Orobanchen, die für ein Herbarium getrocknet werden, sind später immer schwierig zu bestimmen, weil die Farbe verschwindet und dadurch eine Diagnose noch schwerer wird als an der lebenden Pflanze.

2.9 WIRTSPFLANZEN

Viele Arten befallen nur eine Wirtspflanze, andere dagegen parasitieren auf mehreren Pflanzenarten, einige sogar auf mindestens 50 verschiedenen Wirtspflanzen. Umgekehrt sind einige Pflanzen nur Wirt für eine *Orobanche*-Art, dagegen können auf anderen Wirtspflanzen mehrere Arten parasitieren. So schmarotzen ziemlich viele Arten auf *Artemisia campestris* (*Orobanche arenaria, artemisiae-campestris, O. cernua, O. coerulescens* und *O. purpurea* var. *bohemica*). Im Laufe der Zeit hat man auch viele Arten zu unrecht als Wirtspflanze bezeichnet. Die oberirdischen Teile einer Wirtspflanze können oft mehrere Meter von der Sommerwurz entfernt sein. Auf der Suche nach *Orobanche salviae* in der Umgebung von Berchtesgaden in den Bayerischen Alpen (Südostdeutschland) fanden sich inmitten von hunderten Exemplaren des Klebrigen Salbeis (*Salvia glutinosa*), der Wirtspflanze von *O. salviae*, etwa ein Dutzend Schmarotzer. Nach Bestimmung der Art stellte sich heraus, daß es sich hier um *O. flava* handelte, die auf Pestwurz (*Petasites paradoxus*) parasitiert. In der direkten Umgebung war aber nur *Salvia glutinosa* vorhanden. Deswegen wurde eine Pflanze ausgegraben. Die Wurzeln der *O. flava* waren tatsächlich mit Pestwurz verbunden. Der Wirt (Pestwurz) stand einige Meter von *O. flava* entfernt. Eine fast ähnliche Situation wurde in der Nähe von Asbach (Hessisches Bergland, Deutschland) beobachtet. In dieser Gegend sind einige Bestände von *O. bartlingii* vorhanden, die auf *Seseli libanotis* parasitiert. Am gleichen Standort, direkt in der Nähe von *O. bartlingii*, wächst auch *Peucedanum cervaria* (Wirtspflanze von *O. alsatica*) und *Laserpitium latifolium* (Wirtspflanze von *O. alsatica* subsp. *mayeri*).

In diesem Buch sind nur Wirtspflanzen aufgenommen worden, die hauptsächlich in der Natur in Mittel- und Nordeuropa vorkommen. Weil einige Arten auf vielen Wirtspflanzen schmarotzen, besteht die Möglichkeit, daß nicht alle Wirtspflanzen bei jeder Art aufgeführt sind. So wurde *Orobanche purpurea* auch schon schmarotzend auf *Tagetes* spec. beobachtet (Swart, 1972) und *O. minor* auf *Pelargonium* spec. und *Nerium oleander* (Dijkstra, 1969).

seed is infected with the right root fungus in the soil). *Orobanche* seeds are very small and light: approximately 0.001 mg, which is roughly equivalent to the weight of the lightest orchid seeds (Gilli, 1966). Because the seeds are so smal, they are very easily washed down into the deeper layers of the soil. The seeds are dispersed by the wind. As the endosperm contains oil, the germination capacity of many species is maintained for 10-12 years.

2.8 MORPHOLOGY

Most species have a single inflorescence. A few other species, like *Orobanche ramosa*, which also grows in central Europe, usually have branched stems and multiple inflorescences. Normally, several plants grow together; a group of 78 specimens of *O. variegata* has been observed. The inflorescence is usually a dense spike, but in some species, like *O. purpurea* and *O. hederae*, it is lax, with two thirds of the stem covered with flowers.

When broomrapes are found in the field, it is important to determine the colour of the stigma (-lobes) and the corolla-tube, but especially to identify the host. However, the colour of the stigma and of the flower may sometimes deviate from the typical colours. After *Orobanche* species have been pressed for a herbarium, they are difficult to identify, because colours will have disappeared, making identification even harder than from the living material.

2.9 HOST PLANTS

Many species infect only a single type of host plant, others parasitize a range of hosts; some even on at least 50 different hosts. On the other hand, some plants are host to one *Orobanche* species only, while others may have several *Orobanche* species as parasites. Thus, a fairly large number of species are parasitic on *Artemisia campestris* (*Orobanche arenaria, artemisiae-campestris, O. cernua, O. coerulescens* and *O. purpurea* var. *bohemica*). Over the years, many plant species have been wrongly identified as host plants. The above ground parts of a host plant may well be several metres away from the broomrape. Looking for *Orobanche salviae* in the vicinity of Berchtesgaden in the Bayerische Alpen in south-eastern Germany, about a dozen parasites were found amidst hundreds of sticky sage (*Salvia glutinosa*) plants, the host to *Orobanche salviae*. After identification of the *Orobanche* species it turned out to be *O. flava*, which should be parasitic on butterbur (*Petasides paradoxus*). Only *Salvia glutinosa* was found growing in the immediate vicinity. Therefore a plant was dug up. The roots of the parasite were indeed found to be attached to a specimen of butterbur, growing several metres away. A similar situation was observed near Asbach (Hessisches Bergland, Germany). In this area a few groups of *Orobanche bartlingii* were found growing on *Seseli libanotis*. *Peucedanum cervaria* (host to *O. alsatica*) and *Laserpitium latifolium* (host to *O. alsatica* subsp. *mayeri*) were growing in the same location, close to *O. bartlingii*.

The present book includes only those host plants which grow mainly in central and northern Europe. As several species are parasitic on many hosts, it is possible that not all hosts are listed for each *Orobanche* species. Thus, *Orobanche purpurea* has also been found parasitizing on *Tagetes* spec. (Swart, 1972) and *O. minor* on *Pelargonium* spec. and *Nerium oleander* (Dijkstra, 1969).

2.10 HABITAT, ECOLOGY

Most broomrapes grow in sunny, dry and warm places, preferably in arid and semi-arid grassland, but also in xerothermic (e.g. *Orobanche arenaria* and *O. coerulescens*) and nutrient-poor grassland, usually with a southern exposure. Other species are found frequently in ruderal pastures, in thickets (e.g. *O. bartlingii*), on waysides (especially in the Mediterranean region) and at the edges of arable fields. Very few *Orobanche* species grow in shaded places; examples include *O. salviae*, *O. flava* and *O. hederae*, which grow in moist places at the edges of forests or in open parts of mixed woodland (alongside walking trails). *O. laserpitii-sileris* prefers screes at montane levels. A few species also grow in Alpine meadows (*O. gracilis*, *O. alba*). *Orobanche crenata*, *O. ramosa*, *O. minor* and *O. cumana* are parasitic in large numbers on cultivated plants and are therefore found almost exclusively in agricultural environment in central Europe. In nature, these species are often highly selective in the choice of their host.

Broomrapes grow fast, vigorously and abundantly. After a short flowering period the whole plant quickly turns brown. The withered stems will remain standing for a considerable time. Stems of some species can still be found a year later (like *Neottia nidus-avis* among the orchids), so that many *Orobanche* species can still be identified if the host is still recognizable. This applies, for example, to *Orobanche flava* (on *Petasites paradoxus*), *O. hederae* (on *Hedera helix*), *O. lucorum* (on *Berberis vulgaris*), *O. rapum-genistae* (on *Cytisus scoparius*) and *O. salviae* (on *Salvia glutinosa*).

Many species are unpredictable in their behaviour. They never flower the same way and the number of flowering plants can vary greatly each year. In well-known locations, not a single plant may be found for several years, while in a favourable year they suddenly appear again. This is especially true for *O. pallidiflora*, whose flowering time, moreover, varies from year to year. In the Netherlands, flowering plants have been found as early as June, but also in November.

2.11 DISTRIBUTION

Most species grow in temperate or warm regions of Europe, in the adjacent regions of western and central Asia and in northern Africa. In Europe, the largest numbers of species can be found in the Mediterranean countries. Several *Orobanche* species can be found in America, especially in California, and in the Tropics. Crossing the equator southwards, *Orobanche minor* is found in eastern Africa and New Zealand, *O. cernua* in western Australia and *O. ramosa* subsp. *mutelii* (Schultz) Coutinho in the Cape Province. The plants have probably been introduced by man to these areas (Gilli, 1966).

The present book describes all species (except those growing only in the Mediterranean area) which grow in the following countries and regions: Norway, Sweden, Finland, Ireland, United Kingdom, Denmark, the Netherlands, Belgium, Luxembourg, central France, northern France, Germany, Poland, Estonia, Latvia, Lituania, Belorussia, Ukraine, the Czech Republic, Slovakia, Switzerland, northern Italy, Liechtenstein, Austria, Hungary, Rumania, Moldavia, Slovenia, Croatia and northern Yugoslavia. Many species described in this book also grow in Mediterranean countries.

2.12 THREATS AND CONSERVATION

Nearly all central European *Orobanche* species are more or less seriously endangered. Most European species are widely dis-

VERBREITUNGSKARTE MIT ANGABE DER ABKÜRZUNGEN DER LÄNDER

A	Österreich	I	Italien
AL	Albanien	IRL	Irland
B	Belgien	L	Luxemburg
BA	Bosnien-Herzegowina	LT	Litauen
BG	Bulgarien	LV	Lettland
BY	Weißrussland	MA	Marokko
CH	Schweiz	MD	Moldawien
CY	Zypern	MK	Makedonien
CZ	Tschechische Republik	N	Norwegen
D	Deutschland	NL	Niederlande
DK	Dänemark	P	Portugal
DZ	Algerien	PL	Polen
E	Spanien	RO	Rumänien
EW	Estland	RUS	Russland
F	Frankreich	S	Schweden
FIN	Finnland	SK	Slowakei
FL	Liechtenstein	SLO	Slowenien
GB	Großbritannien	TN	Tunesien
GR	Griechenland	TR	Türkei
H	Ungarn	UA	Ukraine
HR	Kroatien	YU	Jugoslawien

DISTRIBUTION MAP WITH ABBREVIATIONS OF COUNTRIES

A	Austria	I	Italy
AL	Albania	IRL	Ireland
B	Belgium	L	Luxembourg
BA	Bosnia-Hercegovina	LT	Lituania
BG	Bulgaria	LV	Latvia
BY	Belorussia	MA	Morocco
CH	Switzerland	MD	Moldavia
CY	Cyprus	MK	Macedonia
CZ	Czech Republic	N	Norway
D	Germany	NL	the Netherlands
DK	Denmark	P	Portugal
DZ	Algeria	PL	Poland
E	Spain	RO	Rumania
EW	Estonia	RUS	Russia
F	France	S	Sweden
FIN	Finland	SK	Slovakia
FL	Liechtenstein	SLO	Slovenia
GB	United Kingdom	TN	Tunisia
GR	Greece	TR	Turkey
H	Hungary	UA	Ukraine
HR	Croatia	YU	Yugoslavia

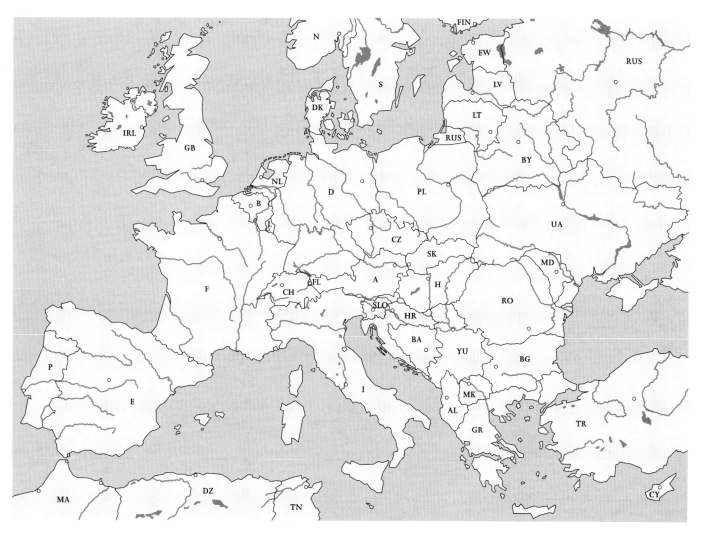

tes Verbreitungsgebiet, sind aber fast überall selten und zählen in den meisten Ländern zu den großen Raritäten. Manche Sommerwurzarten sind endemisch und einige sogar vom Aussterben bedroht. Viele Standorte sind durch Bebauung, Straßenbau und starke landwirtschaftliche Nutzung, durch übermäßige Düngung der Äcker und Wiesen (Eutrophierung) verschwunden. Außerdem vernichtet die Verbuschung der Trockenrasen, die zu ungünstigen Licht- und Temperaturverhältnissen für Wirt und Parasit führt, viele Standorte. Auch das Sammeln von Pflanzen und das Betreten der Standorte kann besonders bei potentiell gefährdeten Arten zu erheblichen Verlusten führen.

So war *Orobanche minor* Anfang des zwanzigsten Jahrhunderts ein großes Problem auf den Äckern verschiedener mitteleuropäischer Länder, wo sie auf verschiedenen Kleearten parasitierte (vgl. unter anderem De Wever, 1918). Durch Kulturmaßnahmen, wie zum Beispiel Wechselwirtschaft und Düngung, ist sie heute an vielen Standorten verschwunden.

Die exakte Kenntnis der Arten und ihrer Fundorte kann zum wirksamen Schutz der Sommerwurzarten beitragen und gezielte Maßnahmen des Naturschutzes ermöglichen. Interessante Vorkommen außerhalb von Naturschutzgebieten sollten unter Schutz gestellt werden. Eine regelmäßige Bestandsüberwachung wertvoller Vorkommen ist notwendig, wobei besonders die vom Aussterben bedrohten Arten beachtet werden sollten. Zur Vermeidung von Trittschäden wäre eine Einzäunung bestimmter Kleinbiotope zu überdenken. Vorrangige Maßnahmen der Pflege müssen neben der Minimierung von Nährstoff- und Schadstoffimmissionen die regelmäßige Entbuschung und der Nährstoffentzug durch Entnahme von Biomasse (Mahd, Schafhut) sein (Pusch & Barthel, 1990).

Zur Zeit sind die *Orobanche*-Arten in fast keinem europäischen Land geschützt abgesehen von Naturschutzgebieten, wo mit ihrer Erhaltung gerechnet werden kann. Deswegen wäre es angebracht wenn alle Sommerwurzarten, mit Ausnahme von Kulturschädlingen, in allen europäischen Ländern geschützt würden. In den Mittelmeerländern sollten zumindest die seltenen Arten unter Naturschutz gestellt werden.

tributed, but are rare in most regions and are considered to be among the great rarities in most countries. Many *Orobanche* species are endemic and a few are on the verge of becoming extinct. Many habitats have been lost because of building activities, road construction and intensive farming, resulting in excessive use of fertilizers on fields and pastures (eutrophication). Many habitats in dry grasslands are further threatened by invading shrubs, resulting in less favourable conditions of light and temperature for both host and parasite. Treading on habitats and collecting plants can also cause major losses of potentially endangered species.

In the early twentieth century, *Orobanche minor* presented serious problems for the arable crops of various central European countries, where it was parasitic on several clover species (see e.g. De Wever, 1918). New cultivation techniques, such as crop rotation and the use of artificial fertilizers, have led to its becoming extinct in most locations.

An exact knowledge of the broomrape species and their habitats may contribute to effective protection and allow specific and efficient measures to be taken. Important stands outside of protected areas should be protected as well. Regular surveys of important habitats, with special attention to endangered species, are essential. Fencing in small areas to prevent damage by treading should be considered. Priority protective measures should include the reduction of the influx of nutrients and pollutants, regular shrub control and reduction of the nutrient concentrations by extraction of bio-mass (mowing, keeping sheep) (Pusch & Barthel, 1990).

At present, broomrapes are unprotected in most European countries, with the exception of those specimens growing in nature reserves, where conservation would appear to be assured. That is why all European countries ought to consider protecting all *Orobanche* species, except those which cause serious damage to crops. In the Mediterranean countries, at least the rare species should be protected.

2.13 ADVENTIVE VORKOMMEN

Einige Arten, die vorwiegend auf Kulturpflanzen in trocken-warmen Ländern im Mittelmeerraum parasitieren, sind in Mitteleuropa höchstwahrscheinlich eingeschleppt worden. Dies gilt für *Orobanche ramosa, O. crenata, O. cumana* und *O. minor*. *O. ramosa* ist inzwischen wieder an vielen Stellen verschwunden, so auch in den Niederlanden, wo die Art früher auf Tabak (*Nicotiana tabacum*) und Hanf (*Cannabis sativa*) parasitierte, aber auch in natürlichen Lebensräumen vorkam. *O. crenata* wurde nur selten in Mitteleuropa beobachtet (unter anderem in Deutschland, Niederösterreich, der Schweiz und Norditalien). *O. cumana* wird in Mitteleuropa vor allem in Sonnenblumenkulturen in Norditalien gefunden. *O. minor* wächst heute in Mitteleuropa vor allem als Wildpflanze in natürlichen Biotopen. Sie parasitiert aber auch auf Kulturpflanzen und kann dann sehr schädlich sein. In letzter Zeit ist sie aber in den meisten Ländern selten geworden.

2.13 ADVENTIVE LOCATIONS

Some species, parasitic mostly on cultivated plants in warm and dry Mediterranean countries, have probably been introduced into central Europe. This applies to *Orobanche ramosa, O. crenata, O. cumana* and *O. minor*. By now, *O. ramosa* has disappeared again from many places, for example from the Netherlands, where the species used to parasitize tobacco (*Nicotiana tabacum*) and hemp (*Cannabis sativa*), but was also found in natural habitats. *O. crenata* was rarely observed in habitats in central Europe (e.g. Germany, Austria (Niederösterreich), Switzerland and northern Italy). *O. cumana* is mostly found in sunflower fields in northern Italy. *O. minor* now usually grows in natural habitats in central Europe, although it also parasitizes cultivated plants, sometimes causing severe damage. It has recently become rare in most countries.

2.14 ERNTESCHÄDEN

In landwirtschaftlichen Kulturen in den Mittelmeer-Ländern und im Nahen Osten haben sich vor allem *Orobanche cumana, O. crenata* und *O. ramosa* und in Mitteleuropa *O. minor* zu regelrechten Landplagen entwickelt; sie verursachen dort große Schäden an verschiedenen Kulturpflanzen. Diese Arten können in Millionen Exemplaren auftreten und in hohem Maße die Ernteerträge beeinträchtigen; in zahlreichen Fällen wurde sogar die Ernte ganze Felder vernichtet. Hierzu genügt oft schon eine Kapsel voller Samen. Die Kulturpflanzen kommen dann nicht mehr zur Blüte und setzen daher auch keine Früchte an. Mit Jäten

2.14 CROP DAMAGE

A few *Orobanche* species in agricultural environments in the Mediterranean countries and the near East (mainly *Orobanche cumana, O. crenata* and *O. ramosa*) as well as *O. minor* in central Europe, have become real plagues; they cause extensive damage to various crops. These species may occur in stands numbering millions of plants and considerably reduce the yield; entire fields have frequently been destroyed. One seed capsule may be enough to destroy the entire crop of a field. The crop plants are unable to flower and therefore cannot produce fruit.

Weeding is hardly a suitable method to eradicate these root parasites. They germinate at the same time as the host plant, adhere to its roots, penetrate into them and 'suck' the young plant dry. The host then rarely flowers; in a few cases the host even dies. In the meantime the parasite has accumulated enough nutrients to flower and bear fruit. So, instead of fields full of tomatoes, tobacco or clover, millions of parasites are the result.

Most species that parasitize crop plants have approximately 20 flowers and each plant produces several hundreds of thousands of seeds. Because each plant can produce so many seeds, a few plants can multiply quickly on fields and seriously affect the harvest. In addition, farmers often contribute to the infection of their own fields. Cattle often feed on *Orobanche* species and their manure is used for fertilizer. Broomrape seeds, however, retain their germinative capacity even during their passage through the digestive tract of these animals. Once the seeds have found their way into the soil of a field, several of them may wait for up to twenty years for the right host plant, before germinating (Hartmann, 1988). The germinative capacity of *Orobanche crenata* lasts for at least 12 years, which is why the same crop should not be grown in the same field within this period (Oliveira-Velloso *et al*, 1994).

Orobanche cumana infects *Helianthus annuus* L. (sunflower) cultures and grows in populations numbering thousands in many places in the Mediterranean countries and in south-eastern Europe. In Bulgaria, this species was found for the first time in 1935, and over the period 1945-1950 it infected more than 80% of the Bulgarian sunflower fields, where up to 86 parasites were sometimes found on a single host plant (Encheva *et al.*, 1994). The species grows abundantly in many sunflower fields in Andalusia, where three parasites may be found infecting one sunflower (Velloso *et al.*, 1993).

Orobanche crenata also causes widespread damage to crops in all Mediterranean countries. It parasitizes *Vicia faba, V. ervilia, Pisum sativum* or *Lens culinaris*. In many countries losses due to this species can amount to 50-80%.

Orobanche ramosa parasitizes various cultivated plants, like tobacco (*Nicotiana tabacum*), tomatoes (*Solanum lycopersicum*), hemp (*Cannabis sativa*) and maize (*Zea mays*), and can cause a reduction of the yield of up to 40%.

Orobanche minor grows not only in Mediterranean countries but also in large parts of northern and central Europe. It is parasitic mostly on red clover (*Trifolium pratense*), but has also been found on sunflowers (Kreeftenberg, 1992). A single *Orobanche minor* plant in a field of *Trifolium pratense* in the Netherlands was sufficient for the Ministry of Agriculture to condemn the entire crop.

Seeds of all these species have been transported from their presumably natural habitat in the Mediterranean region into North America, Asia and Australia, where the plants have been spreading rapidly in arid regions (Weber, 1993).

2.15 CONTROLLING *OROBANCHE* SPECIES IN AGRICULTURAL ENVIRONMENTS

If only a few *Orobanche* plants are found in a field, the easiest way to remove them is by weeding, but collecting the plants by hand is very time-comsuming. This method can only be applied on small fields with a low level of infection.

Another method is to sow later, shortening the time available for *Orobanche* to develop; this may reduce the infection by as much as 90%. Opportunities for applying this method are limited, however, as it results in less time for the development of the crop or other crop losses due to the postponement (Weber, 1993).

den kann oder andere durch die Zeitverschiebung bedingte Ertragseinbußen in Kauf genommen werden müssen (Weber, 1993).
Wechselwirtschaft ist ebenfalls nicht sehr effektiv, weil die Keimfähigkeit der meisten Samen bis zu 12 Jahren betragen kann.
Weiter kann auch Tiefpflügen die Infektion reduzieren, weil die *Orobanche*-Samen dann tiefer in den Boden gelangen. Orobanchen entwickeln sich optimal in einer Bodentiefe bis zu 20 cm. Tiefes Umpflügen bis zu 50 cm kann die Keimung stark reduzieren. Das Problem ist aber, daß viele Böden in den Mittelmeerländern dafür überhaupt nicht geeignet sind.
Gute Ergebnisse wurden auch erreicht durch Vorkeimung, die Anzucht resistenter Kulturpflanzen, Düngung, Anwendung von Chemikalien (Herbiziden), Solarisation (die Temperatur auf dem Lande stark erhöhen) und durch das Einsetzen bestimmter Insekten und Anbau bioherbiziden Pilzarten. So gehört *Orobanche cumana* zu den Orobanchen, die sich am schwierigsten bekämpfen lassen. Viele Mittel zur Reduzierung dieser Art waren vergebens. Bekämpfung mit biologischen Mitteln, zum Beispiel mit *Fusarium oxysporum* subsp. *orthroceras* haben sich dagegen als effektiv erwiesen. Die Anwendung dieser Pilzart hat die Zahl der Parasiten um 90% zurückgedrängt (Bedi, 1994).
Die Erhöhung der Resistenz oder Toleranz gegenüber Orobanchen wurde überraschenderweise in einigen Fällen äußerst erfolgreich durch die Züchtung von Wirtsvarietäten erzielt (genetische Veränderungen). Die resistenten oder toleranten Wirtsvarietäten weisen aber oftmals eine schlechtere Fruchtqualität auf und sind weniger ertragreich. Außerdem bleibt abzuwarten, wie lange es dauern wird, bis sich der Parasit an "seinen neuen Partner" gewöhnt hat (Weber, 1993).
Seit etwa einem Jahrhundert beschäftigt sich die Wissenschaft mit der Bekämpfung der Orobanchen in landwirtschaftlichen Kulturen. Etwa jedes dritte Jahr findet ein Internationales Symposium über parasitäre Blütenpflanzen statt. Das letzte Symposium wurde im November 1993 in Amsterdam (die Niederlande) abgehalten.

2.16 LITERATUR

Einige wichtige Monographien wurden von Vaucher (1827), Schultz (1829), Koch (1887), Guimaraes (1904), Beck von Mannagetta (1890) und von Beck-Mannagetta (1930) verfaßt. Andere wichtige Werke stammen von Gilli, der die Orobanchen für Hegi (1966) bearbeitet hat; von Chater & Webb (*Flora Europaea*, 1972) und von Pusch & Barthel (*Merkmale und Verbreitung der Gattung Orobanche L. in den östlichen Bundesländern Deutschlands*, 1992). Von jedem europäischen Land sind mehrere (Exkursions-)Floren erschienen, die alle *Orobanche*-Arten des betreffenden Landes mit einem Bestimmungsschlüssel enthalten.
Im Laufe der Zeit sind auch viele Artikel über Orobanchen in verschiedenen botanischen Zeitschriften erschienen.
Über die Bekämpfung der Sommerwurzgewächse in landwirtschaftlichen Kulturen wurde bis jetzt in drei Heften, *Biology and control of Orobanche* (Ter Borg, 1986), *Progress in Orobanche research* (Wegmann & Musselman, 1991) und *Biology and management of Orobanche* (Pieterse *et al.*, 1994) berichtet.

Crop rotation is not very effective, since the germinative capacity of the parasite's seeds can remain intact for up to 12 years.
Deep ploughing can also reduce the infection, because the *Orobanche* seeds are pushed down deeper into the soil. *Orobanche* seeds reach their optimum development at depths of 20 cm. Ploughing to a depth of 50 cm will greatly reduce the germination of the seeds. A problem is that many types of soil in the Mediterranean countries are not suitable for deep ploughing at all.
Good results have also been obtained with pregerminated seed, the cultivation of resistant plants, adding fertilizer, the use of chemicals (herbicides), solarization (increasing the temperature of the soil), and the use of certain insects or bio-herbicidal fungi. *Orobanche cumana* is one of the most difficult species to destroy. Many control measures have failed, but the use of biological methods, like *Fusarium oxysporum* subsp. *orthroceras*, has proved to be effective. Use of this fungus has helped to reduce the parasite by 90% (Bedi, 1994).
Selective breeding of (genetically altered) host plant varieties has in some cases led to a surprising increase in their resistance and tolerance to *Orobanche*. However, this has often also resulted in poorer fruit quality and yield for the resistant or tolerant host plant varieties. Another question is how long the parasite will need to adapt to the "new partner" (Weber, 1993).
For about a century now, science has attempted to control *Orobanche* in agriculture. An international symposium on parasitic flowering plants is organized approximately every three years. The last symposium was held in November 1993 in Amsterdam (the Netherlands).

2.16 BIBLIOGRAPHY

A few important monographs have been written by Vaucher (1827), Schultz (1829), Koch (1887), Guimaraes (1904), Beck von Mannagetta (1890) and by Beck-Mannagetta (1930). Other important books are those by Gilli, who treated the *Orobanche* species for Hegi (1966); by Chater & Webb (*Flora Europaea*, 1972) and by Pusch & Barthel (*Merkmale und Verbreitung der Gattung Orobanche L. in den östlichen Bundesländern Deutschlands*, 1992). Each European country has published one or more (excursion) floras, incorporating all *Orobanche* species found in that particular country, with keys.
Over the years, many articles on *Orobanche* species have been published in several botanic journals.
Up until now, three booklets on the control of broomrapes in agriculture have appeared, entitled *Biology and control of Orobanche* (Ter Borg, 1986), *Progress in Orobanche research* (Wegmann & Musselman, 1991) and *Biology and management of Orobanche* (Pieterse *et al.*, 1994).

SPEZIELLER TEIL   SPECIFIC PART

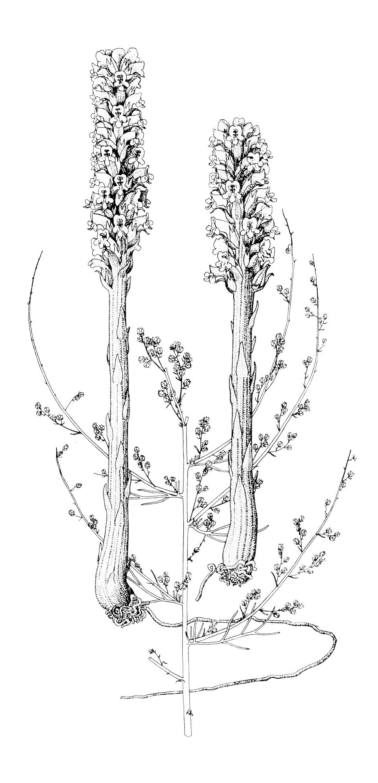

Orobanche coerulescens mit Wirtspflanze (*Artemisia campestris*)   *Orobanche coerulescens* with host (*Artemisia campestris*)

| Bestimmung der mittel- und nordeuropäischen Orobanche-Arten | Identification of central and northern European Orobanche species |

3.1 **BLÜTENBAU**
STRUCTURE OF THE FLOWER
BLOEMBOUW

A Vorblätter (nur Arten der Sektion *Trionychon*)
Bracteolus (species of the section *Trionychon* only)
Steelblaadjes (alleen bij de soorten van de sectie *Trionychon*)

B Tragblatt
Bract
Schutblad

C Kelchhälften
Calyx segments
Kelkhelften

D Blütenkrone (Blumenkrone)
Corolla
Bloemkroon

E Kronröhre
Corolla-tube
Kroonbuis

F Rückenlinie der Blütenkrone (Krümmung der Kronröhre)
Dorsal line of the corolla (curve of the corolla-tube)
Ruglijn van de bloemkroon (kromming van de kroonbuis)

G Oberlippe der Blütenkrone
Upper lip of corolla
Bovenlip van de bloemkroon

H Unterlippe der Blütenkrone
Lower lip of corolla
Onderlip van de bloemkroon

I Zipfel der Oberlippe
Margin of the upper lip
Zoom van de onderlip

J Ansatzstelle der Staubblätter über dem Grund der Kronröhre
Insertion of stamens above base of the corolla-tube
Inplanting van de meeldraden op de kroonbuis

K Staubblätter (Staubfäden und Staubbeutel)
Stamens (filaments and anthers)
Meeldraden (helmdraden en helmknoppen)

L Staubfäden
Filaments
Helmdraden

M Staubbeutel
Anthers
Helmknoppen

N Stempel (Griffel, Narbe und Fruchtknoten)
Carpel (style, stigma and ovary)
Stamper (stijl, stempel en vruchtbeginsel)

O Griffel
Style
Stijl

P Narbe
Stigma
Stempel

R Fruchtknoten
Capsule (ovary)
Vruchtbeginsel

S Drüsenhaare
Glandular hairs
Klierharen

T Saum
Edge
Zoom

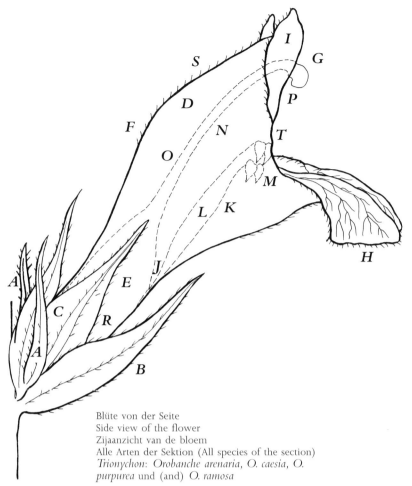

Blüte von der Seite
Side view of the flower
Zijaanzicht van de bloem
Alle Arten der Sektion (All species of the section)
Trionychon: *Orobanche arenaria*, *O. caesia*, *O. purpurea* und (and) *O. ramosa*

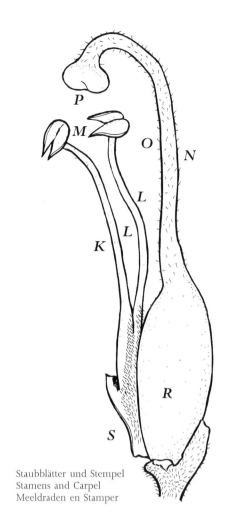

Staubblätter und Stempel
Stamens and Carpel
Meeldraden en Stamper

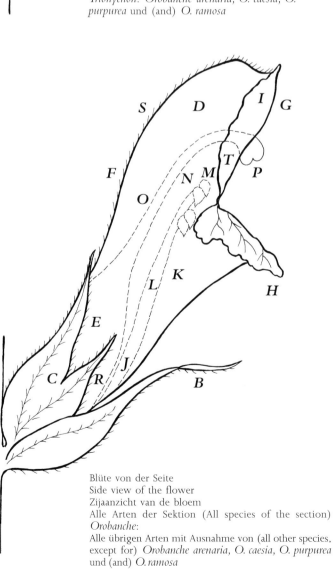

Blüte von der Seite
Side view of the flower
Zijaanzicht van de bloem
Alle Arten der Sektion (All species of the section)
Orobanche:
Alle übrigen Arten mit Ausnahme von (all other species, except for) *Orobanche arenaria*, *O. caesia*, *O. purpurea* und (and) *O. ramosa*

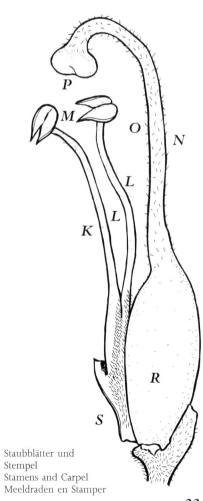

Staubblätter und
Stempel
Stamens and Carpel
Meeldraden en Stamper

3.2 ÜBERSICHT VON FACHAUSDRÜCKEN
GLOSSARY OF TERMS
OVERZICHT VAN VAKTERMEN

Ähre / spike / aar
Angedrückt / appressed / aangedrukt
Aufgeblasen / inflated / opgeblazen
Aufrecht / erect / rechtopstaand
Ausgerandet / emarginate / uitgerand
Bastard / hybrid / bastaard (hybride)
Befruchtung / fertilization / bevruchting
Blatt / leaf / blad
Blumenstiel / pedicel / bloemsteel
Blüte / flower / bloem
Blütenfarbe / colour of flowers / kleur van de bloemen
Blütenstand / inflorescence / bloeiwijze
Blütenstaub / pollen / stuifmeelkorrel
Blütezeit / flowering time / bloeitijd
Chromosomenzahl / chromosome number / chromosomen
Dachziegelig / imbricate / dakpansgewijs
Dreieckig / triangular / driehoekig
Drüsenhaar / glandular hair / klierhaar
Drüsenhaarig / glandular-pubescent / beklierd (klierharig)
Einheimisch / native / inheems
Endemisch / endemic / endeem
Endosperm / endosperm / kiemwit
Fadenförmig / filiform / draadvormig
Filzig / tomentose / viltig
Frucht / fruit / vrucht
Fruchtbar / fertile / vruchtbaar
Gebogen / curved / gebogen
Gedreht / convolute / gedraaid
Gefranst (gewimpert) / ciliate / gewimperd
Gefurcht / sulcate / gegroefd
Gekielt / keeled / gekield
Geknickt / inflected / geknikt
Gelappt / lobate (lobed) / gelobd
Gerippt / ribbed / geribd
Geruch / smell / reuk
Gesägt / saw-toothed / gezaagd
Geschuppt / scaled (lepidoid) / geschubd
Gewimpert / fringed / gewimperd
Gezähnelt / crenate / gekarteld
Gezähnt / dentate (denticulate) / getand
Glockenförmig / campanulate / klokvormig
Haarig / hairy / behaard
Herabgeschlagen / deflexed / teruggeslagen
Infloreszens / inflorescence / bloeiwijze
Kahl / glabrous (glabrescent) / kaal
Knospe / eye / knop
Kräftig / stout, robust / krachtig
Kurzhaarig / hirsute / kortharig
Langhaarig / villous, long-haired / langharig
Länglich / oblong / langwerpig
Lanzettförmig / lanceolate / lancetvormig
Linear / linear / lijnvormig
Lippenförmig / lipped / lippenvormig
Locker / lax / los
Nerv / vein / nerf
Nerven / veins / nerven
Netzartig / reticulate / netvormig
Nickend / nodding / knikkend
Niedergebogen / deflexed / neergebogen
Nierenförmig / reniform (kidney-shaped) / niervormig
Oval / oval (ovate) / eirond (ovaal)
Parallelnervig / parallel-nerved / parallelnervig
Pfriemförmig / awl-shaped / priemvormig
Rauhwollig / villous / ruwharig
Röhre / tube / buis
Röhrenförmig / tubular / buisvormig
Rund / orbicular / rond
Sanfthaarig / pubescent / zachtharig
Saum / margin / zoom, rand
Schlund / throat / keel
Schuppen / scale leaves / schubben
Schwach / weak / zwak

Sitzend / sessile / zittend
Spaltig / segmented / gesegmenteerd
Spärlich / sparsely / spaarzaam
Spinnwebartig / arachnoid / spinnewebharig
Spitz / acute / spits
Stengel / stem / stengel
Stinkend / foetid / stinkend
Überlaufen / tinged / aangelopen
Unauffällig / indistinct / onopvallend
Unverzweigt / simple / onvertakt
Verbreitung / area / areaal
Verwachsen / connate, fused / vergroeid
Verzweigt / branched / vertakt
Vorkommen / habitat / biotoop
Winzig / minute / zeer klein
Wirt / host / waardplant
Wohlriechend / fragant / welriekend
Wollig / woolly / wollig
Zugespitzt / acuminate (elongated, pointed)) / toegespitst
Zungenförmig / lingulate (linguiform) / tongvormig
Zurückgebogen / recurved / teruggebogen
Zusammengezogen / constricted / samengetrokken
Zweizähnig / bidentate / tweetandig

3.3 DIE ARTEN/THE SPECIES

SEKTION/SECTION **TRIONYCHON** Wallroth 1822
1. *Orobanche arenaria*   Borkhausen in Römer's, Neues Mag. Bot. 1: 6 (1794)
2. *Orobanche caesia*   Reichenbach, Pl. Crit. 7: 48 (1829)
3. *Orobanche purpurea*   Jacquin, Enum. Stirp. Vindob. 108: 252 (1762)
4. *Orobanche ramosa*   Linnaeus, Sp. Pl. 633 (1753)

SEKTION/SECTION **OROBANCHE** Linnaeus 1753
5. *Orobanche alba*   Stephan ex Willdenow, Sp. Pl. 3: 350 (1800)
6. *Orobanche alsatica*   Kirschleger, Prodr. Fl. Alsace 109 (1836)
7. *Orobanche alsatica* subsp. *mayeri*   (Suessenguth et Ronniger), C.A.J. Kreutz in Kreutz, C.A.J., Orobanche 56 (1995)
8. *Orobanche amethystea*   Thuillier, Fl. Paris ed. 2: 317 (1797)
9. *Orobanche artemisiae-campestris*   Vaucher ex Gaudin, Fl. Helv, 4: 179 (1829)
10. *Orobanche bartlingii*   Grisebach, Spicil. Fl. Rumel. 2: 57-58 (1844)
11. *Orobanche caryophyllacea*   Smith, Trans. Linn. Soc. London 4: 169 (1798)
12. *Orobanche cernua*   Loefling, Iter. Hisp. 152 (1758)
13. *Orobanche coerulescens*   Stephan ex Willdenow, Sp. Pl. 3: 349 (1800)
14. *Orobanche crenata*   Forskal, Fl. Aegypt. Arab. 113 (1775)
15. *Orobanche cumana*   Wallroth, Orob. Gen. 58 (1825)
16. *Orobanche elatior*   Sutton, Trans. Linn. Soc. London 4: 178 (1797)
17. *Orobanche flava*   Martius ex F.W. Schultz, Beitr. Kenntn. Deutsch. Orob. 9 (1829)
18. *Orobanche gracilis*   Smith, Trans. Linn. Soc. London 4: 172 (1798)
19. *Orobanche hederae*   Duby, Bot. Gall. 1: 350 (1828)
20. *Orobanche laserpitii-sileris*   Reuter ex Jordan, Observ. Pl. Crit. 3: 223 (1846)
21. *Orobanche lucorum*   A. Braun ex Koch in: Röhling, Deutschl. Fl. ed. 3, 4: 456 (1833)
22. *Orobanche lutea*   Baumgarten, Enum. Stirp. Transsylv. 2: 215 (1816)
23. *Orobanche minor*   J.E. Smith, Sowerby, Engl. Bot. 6: 422 (1797)
24. *Orobanche pallidiflora*   Wimmer et Grabowski, Fl. Siles. 2, 1: 233 (1829)
25. *Orobanche picridis*   F.W. Schultz ex Koch in Röhrling, Deutschl. Fl. ed. 3, 4: 453 (1833)
26. *Orobanche rapum-genistae*   Thuillier, Fl. Paris ed. 2, 1: 317 (1799)
27. *Orobanche reticulata*   Wallroth, Orob. Gen. 42 (1825)
28. *Orobanche salviae*   F.W. Schultz ex Koch in: Röhrling, Deutschl. Fl. ed. 3, 4: 458 (1833)
29. *Orobanche teucrii*   Holandre, Fl. de la Moselle ed. 1, 2: 322 (1829)
30. *Orobanche variegata*   Wallroth, Orob. Gen. 40 (1825)

3.4 SUGGESTIONS FOR IDENTIFICATION

All *Orobanche* species can be troublesome to identify, as they vary widely in size and colours. Young, sprouting plants and wilting flowers frequently show atypical colours, especially the stigma. It is recommended to use only fully grown plants for identification, in order to prevent erroneous identifications. It is essential to determine the colours of the corolla and stigma of the living plant and to identify the probable host. Identification often remains difficult, however, as it is hardly possible to include all extreme forms of variable species in a key, without the latter becoming too cumbersome to use.

Identifying the host may not be easy either, as it does not always grow in the immediate vicinity of the broomrape, making it difficult to establish the actual host. As most *Orobanche* species are rare, digging up the plant should definitely be avoided for conservation reasons.

3.5 KEY TO *OROBANCHE* SPECIES IN CENTRAL AND NORTHERN EUROPE

1
Calyx tubular-campanulate, 4-, rarely 5-dentate, entire. Each flower has one bract and two opposite bracteoles, frequently fused to the calyx. Corolla blue or violet, at least above (very rarely white) » 2

1*
Calyx consisting of two bifid or entire halves, joined together below, often at the front, rarely also at the back. Each flower has only one bract, but no bracteoles » 6

2
Stem usually branched, slender, sparsely scaly with small scale leaves, pale yellow. Flowers usually only 10-12 mm, ultimately up to 17 mm, pale yellowish, with blue, violet or white margins. Mostly on *Lamium* species and on various crop plants like *Cannabis sativa* and *Nicotiana tabacum* » 4.4 *O. ramosa* L.

2*
Stem unbranched, in exceptional cases branched. Flowers larger » 3

3
Anthers hairy (woolly). Flowers 26-35 mm. Especially on *Artemisia campestris* » 4.1 *O. arenaria* Borkh.

3*
Anthers glabrous or rarely downy at the base » 4

4
Stem woolly with white hairs, especially above, 10-30 cm tall. Flowers only 20-25 mm. On *Artemisia* species (mostly *A. austriaca* and *A. pontica*). Only in the eastern part of central Europe (mainly eastern part of Austria and Hungary) » 4.2 *O. caesia* Rchb.

4*
Stem glandular-pubescent, but not woolly with white hairs, 15-60 cm tall. Flowers mostly 25-32 mm, rarely smaller » 5

5
On *Achillea* species and on other *Asteraceae* species » 4.3 *O. purpurea* Jacq.

5*
Pflanze meistens sehr kräftig und in allen Teilen stärker gefärbt (dunkelviolett), Blütenstand lang, viel- und dichtblütig, Blütenkrone 20 bis 25 mm lang, Staubbeutel kahl. Nur auf *Artemisia campestris*. Hauptsächlich in Osteuropa » 4.3 *O. purpurea* Jacq. var. *bohemica* (Celak.) Beck

6
Blumenkrone blauviolett oder hellblau, gegen den Grund zu weißlich. Kronröhre unter der Ansatzstelle der Staubblätter bauchig erweitert, darüber eingeschnürt und gegen den Saum zu erweitert. Staubblätter 4 bis 6 mm hoch über dem Grund der Kronröhre eingefügt. Narbe weißlich. Auf *Artemisia*-Arten oder *Helianthus annuus* » 7

6*
Blumenkrone nicht zur Gänze blauviolett. Kronröhre unter der Ansatzstelle der Staubblätter nicht erweitert, darüber nicht eingeschnürt. Staubblätter meist im unteren Drittel der Kronröhre eingefügt. Narbe gelblich, bräunlich oder violett » 9

7
Oberer Teil des Stengels, Tragblätter und Kelche weißwollig. Blumenkrone meist 15 bis 17 mm lang » 4.13 *O. coerulescens* Steph. ex Willd.

7*
Pflanze drüsig behaart, aber nicht weißwollig. Blumenkrone 12 bis 15 mm lang » 8

8
Pflanze kräftig. Blütenstand dicht- und reichblütig, Blüten vor allem gegen den Saum zu (blau)violett gefärbt und meistens kleiner als 15 mm. Vor allem im mediterranen Gebiet. Auf *Artemisia*-Arten » 4.12 *O. cernua* Loefl.

8*
Pflanze schlank. Blütenstand sehr locker und lang. Blüten meistens größer als 15 mm. Fast ausschließlich in Sonnenblumenkulturen (*Helianthus annuus*) Mittel- und Südeuropas » 4.15 *O. cumana* Wallr.

9
Rückenlinie der Blumenkrone (von der Seite gesehen) gerade oder schwach konkav oder schwach konvex und über der Oberlippe plötzlich winkelig abwärts, an der Spitze oft wieder aufwärts gebogen » 10

9*
Rückenlinie der Blumenkrone vom Grund bis zur Spitze ziemlich gleichmäßig gekrümmt oder zumindest nicht plötzlich über der Oberlippe abwärts gebogen » 23

10
Blumenkrone besonders an der Oberlippe mit roten oder purpurnen Drüsenpunkten besetzt, weiß oder gelblich, gegen den Saum zu violett oder rötlich » 11

10*
Blumenkrone nur mit hellen Drüsenhaaren » 13

11
Kelchblätter deutlich nervig, getrocknet meist braun. Staubblätter am Grunde behaart, oben wie der Griffel mit meist reichlichen Drüsenhaaren besetzt. Hauptsächlich auf *Lamiaceae*-Arten » 4.5 *O. alba* Steph. ex Willd.

5*
Plant usually very stout and more intensely coloured in all parts (dark violet), inflorescence long and dense, with numerous flowers, corolla 20-25 mm, anthers glabrous. Only on *Artemisia campestris*. Mainly in eastern Europe » 4.3 *O. pupurea* Jacq. var. *bohemica* (Celak.) Beck

6
Corolla blue-violet or bright blue, whitish below. Corolla-tube inflated below insertion of the stamens, contracted above and widening towards the margin. Stamens inserted 4-6 mm above the base of the corolla-tube. Stigma whitish. On *Artemisia* species or *Helianthus annuus* » 7

6*
Corolla not entirely blue-violet. Corolla-tube not inflated below the insertion of the stamens, not contracted above. Stamens usually inserted in the lower third of the corolla-tube. Stigma yellowish, brownish or violet » 9

7
Upper part of stem, bracts and calyx woolly with white hairs. Calyx usually 15-17 mm » 4.13 *O. coerulescens* Steph. ex Willd.

7*
Plant glandular-pubescent, but not woolly with white hairs. Calyx 12-15 mm » 8

8
Plant stout. Inflorescence dense with numerous flowers. Flowers blue or blue-violet, especially towards the margin, and usually smaller than 15 mm. Mainly in the Mediterranean area. On *Artemisia* species » 4.12 *O. cernua* Loefl.

8*
Plant slender. Inflorescence very lax and elongated. Flowers usually larger than 15 mm. Almost exclusively in sunflower cultures (*Helianthus annuus*) of central and southern Europe » 4.15 *O. cumana* Wallr.

9
Dorsal line of the corolla (as seen from side) straight, slightly concave or slightly convex, suddenly turning down almost perpendicularly over the upper lip, frequently turning upwards again at the apex » 10

9*
Dorsal line of the corolla regularly curved throughout or at least not turning downwards over the upper lip » 23

10
Corolla speckled with red or purple glands, especially on the upper lip, white or yellowish, violet or reddish near the margin » 11

10*
Corolla with pale glandular hairs only » 13

11
Calyx lips clearly veined, brown when dried. Stamens are hairy below, richly glandular-pubescent above, as is the style. Mainly on *Lamiaceae* species » 4.5 *O. alba* Steph. ex Willd.

11*
Kelchblätter undeutlich nervig, getrocknet meist schwärzlich. Kelchhälften fast immer ungeteilt. Griffel spärlich drüsenhaarig. Auf *Asteraceae*- und *Dipsacaceae*-Arten » 12

12
Blumenkrone mehr oder weniger intensiv violett oder purperviolett, an der Basis gelblich gefärbt und dicht mit dunkelvioletten Drüsenhaaren besetzt. Staubblätter an der Basis kahl oder spärlich behaart, in der Mitte fast kahl und im oberen Teil drüsenhaarig. Vor allem in den höheren Lagen der Alpen in Unkrautgesellschaften verbreitet » 4.27 *O. reticulata* Wallr.

12*
Blumenkrone weißlich oder gelblich, gegen den Saum zu schwach lila, mit spärlichen dunklen Drüsenhaaren besetzt. Staubblätter oben spärlich drüsenhaarig bis kahl. Nur in den niederen Lagen auf *Cirsium arvense, C. oleraceum, Carduus acanthoides* und *C. crispus* (hauptsächlich in Ruderalvegetationen) » 4.24 *O. pallidiflora* Wimm. et Grab.

13
Blumenkrone weiß oder gelblich, meist gegen den Saum zu mindestens an den Nerven violett, bläulich oder rötlich, 10 bis 22 mm lang » 14

13*
Blumenkrone braun oder braunviolett, nur am Grunde weißlich oder gelblich, selten die ganze Blumenkrone gelb, weißlich oder lila, meist 20 bis 35 mm lang » 20

14
Blüten abstehend bis aufrecht-abstehend, ca. 5 bis 8 mm breit, purpurn überlaufen, drüsenhaarig bis fast kahl. Auf verschiedenen Wirtspflanzen » 15

14*
Pflanze violett gefärbt, bis zu 60 cm hoch. Blütenstand im unteren Teil locker- und im oberen Teil dichtblütig. Narbe rotbraun bis violett. Auf verschiedenen Wirtspflanzen. Verbreitung vor allem in Ost-Anglia und Surrey (England) » 16

15
Mittellappen der Unterlippe der Blütenkrone nierenförmig und größer als die beiden Seitenlappen. Blüten etwa 6 bis 10 mm groß, unterste Blüten meistens gestielt. Die zwei halbkugeligen Lappen der Narbe teilweise miteinander verbunden. Auf *Daucus carota*, selten auf *Plantago coronopus* und *Ononis repens*. Nur in Südengland, auf den Kanal-Inseln und Westfrankreich » 4.23 *O. minor* Sm. var. *maritima* (Pugsley) Rumsey & Jury

15*
Blumenkrone 10 bis 17, meist 15 mm lang, gegen den Schlund wenig erweitert, Lappen der Oberlippe vorgestreckt oder abstehend, Staubblätter 2 bis 3 mm hoch über dem Grund der Kronröhre eingefügt. Verbreitung fast ganz Europa. Vorwiegend auf *Trifolium*-Arten » 4.23 *O. minor* Sm. s.l.

16
Blüten bleich violett gefärbt und spärlich mit Drüsenharen besetzt. Blütenkrone 12 bis 18 mm lang und 3,5 bis 5 mm breit. Staubblätter an der Basis dicht behaart. Auf *Asteraceae*-Arten, hauptsächlich auf *Crepis virens, Hypochoeris radicata, Tripleurospermum inodorum, Carduus nutans* und *Senecio greyii* » 4.23 *O. minor* Sm. var. *compositarum* Pugsley

16*
Blumenkrone meist größer, 10 bis 22 mm lang. Staubblätter 3 bis 5 mm hoch über dem Grund der Kronröhre eingefügt. Nicht auf *Fabaceae*-Arten » 17

11*
Calyx indistinctly veined, blackish when dried. Calyx lips usually entire. Style sparsely glandular-pubescent. On *Asteraceae* and *Dipsacaceae* species » 12

12
Corolla more or less brightly violet or purple-violet, yellowish below, densely glandular-pubescent with dark violet glandular hairs below. Stamens sparsely hairy or glabrous below, almost glabrous in the middle and glandular-pubescent above. Mainly in higher altitudes in the Alps in herbaceous vegetation » 4.27 *O. reticulata* Wallr.

12*
Corolla whitish or yellowish, tinged with violet towards the margin, sparsely glandular-pubescent with dark hairs. Stamens sparsely glandular-pubescent to glabrous above. Only in low altitudes on *Cirsium arvense, C. oleraceum, Carduus acanthoides* and *C. crispus* (mainly in ruderal vegetations) » 4.24 *O. pallidiflora* Wimm. et Grab.

13
Corolla white or yellowish; violet, bluish or reddish towards the margin, at least at veins, 10-22 mm » 14

13*
Corolla brown or brown-violet, whitish or yellowish only at base, rarely with the entire corolla yellow, whitish or violet, mostly 20-35 mm » 20

14
Flowers spreading to erecto-patent, approximately 5-8 mm wide, tinged with purple, glandular-pubescent to almost glabrous. On various host plants » 15

14*
Plant violet, up to 60 cm tall. Inflorescence lax below, dense above. Stigma red-brown to violet. On various host plants. Mainly in East-Anglia and Surrey (United Kingdom) » 16

15
Middle lobe of the lower lip of the corolla reniform and larger than other lobes. Flowers approximately 6-10 mm, lower flowers usually pediculate. Hemispherical lobes of stigma partly fused. On *Daucus carota*, rarely on *Plantago coronopus* and *Ononis repens*. Only in southern England, on the Channel Islands and in western France. » 4.23 *O. minor* Sm. var. *maritima* (Pugsley) Rumsey & Jury

15*
Corolla 10-17 mm, mostly 15 mm, slightly inflated near the throat, lobes of the upper lip spreading or erecto-patent, stamens inserted 2-3 mm above the base of the corolla-tube. Throughout most of Europe. Mainly on *Trifolium* species » 4.23 *O. minor* Sm. s.l.

16
Flower pale violet and sparsely glandular-pubescent. Corolla 12-18 mm and 3.5-5 mm wide. Stamens densely glandular-pubescent below. On *Asteraceae* species, mainly on *Crepis virens, Hypochoeris radicata, Tripleurospermum inodorum, Carduus nutans* and *Senecio greyii* » 4.23 *O. minor* Sm. var. *compositarum* Pugsley

16*
Corolla usually larger, 10-22 mm. Stamens inserted 3-5 mm above the base of the corolla-tube. Not on *Fabaceae* species » 17

17
Blumenkrone unter dem Schlund zusammengezogen, mit einem abstehenden, viel weiteren Saum. Narbe gelb bis orange. Meist auf *Hedera helix* ›› 4.19 *O. hederae* Duby

17*
Blumenkrone nicht unter dem Schlund zusammengezogen. Narbe violett, rot oder rotbraun. Nicht auf *Hedera helix* ›› 18

18
Griffel reichlich drüsenhaarig. Kelchhälften meist bis unter die Mitte zweizähnig, seltener ungeteilt. Nur auf *Artemisia campestris* ›› 4.9 *O. artemisiae-campestris* Vaucher ex Gaudin

18*
Griffel spärlich drüsenhaarig. Kelchhälften höchstens bis zur Mitte zweizähnig oder ungeteilt. Nicht auf *Artemisia*-Arten ›› 19

19
Blüten aufrecht abstehend, später etwas stärker abstehend. Auf *Picris hieracioides* und *Daucus carota* ›› 4.25 *O. picridis* F.W. Schultz ex Koch

19*
Blüte knieförmig gebogen. Vor allem westliches Mittel- und Südeuropa. Nur auf *Eryngium campestre* ›› 4.8 *O. amethystea* Thuill.

20 (13*)
Narbe gelb oder rötlich. Auf *Fabaceae*-Arten ›› 21

20*
Narbe purpurn oder braun, sehr selten gelb. Nicht auf *Fabaceae*-Arten ›› 22

21
Narbe gelb. Blumenkrone braun, unten gelblich, oft mit violetten Nerven, seltener zur Gänze purpurn oder gelblich ›› 4.22 *O. lutea* Baumg.

21*
Narbe zuerst gelb, dann rötlich. Blumenkrone mit Ausnahme der Basis braunpurpurn oder schwärzlich, innen intensiv gefärbt und glänzend. Mittellappen der Unterlippe der Blütenkrone meistens doppelt so groß wie die beiden Seitenlappen. Verbreitung vor allem westliches Mittelmeergebiet bis Norditalien ›› 4.30 *O. variegata* Wallr.

22
Staubblätter nahe dem Grund der Kronröhre eingefügt. Blumenkrone zur Gänze hellbraunviolett oder rötlich, selten gelb. Meist auf *Galium*- und *Asperula*-Arten ›› 4.11 *O. caryophyllacea* Smith

22*
Staubblätter 3 bis 5 mm hoch über dem Grund der Kronröhre eingefügt. Blumenkrone hellbraunlila, am Grund weißlich. Auf *Teucrium*-Arten ›› 4.29 *O. teucrii* Holandre

23 (9)
Blumenkrone außen gelb, an der Oberlippe und an den Nerven oder zur Gänze rot oder purpurn, innen meist glänzend-purpurn und getrocknet schwarzbraun, selten zur Gänze wachsgelb. Narbe meist gelb, blutrot oder purpurn gesäumt ›› 24

23*
Blumenkrone innen nicht glänzend purpurn ›› 25

17
Corolla contracted at the throat, with a much wider, spreading margin. Stigma yellow to orange. Mostly on *Hedera helix* ›› 4.19 *O. hederae* Duby

17*
Corolla not contracted below the throat. Stigma violet, red or red-brown. Not on *Hedera helix* ›› 18

18
Style densely glandular-pubescent. Calyx-segments mostly deeply bifid to below the middle, rarely entire. Only on *Artemisia campestris* ›› 4.9 *O. artemisiae-campestris* Vaucher ex Gaudin

18*
Style sparsely glandular-pubescent. Calyx-segments mostly bifid down to the middle or entire. Not on *Artemisia* species ›› 19

19
Flowers erect to spreading, more clearly so later on. On *Picris hieracioides* and *Daucus carota* ›› 4.25 *O. picridis* F.W. Schultz ex Koch

19*
Flower acutely geniculate. Mainly in western central and southern Europe. Only on *Eryngium campestre* ›› 4.8 *O. amethystea* Thuill.

20 (13*)
Stigma yellow or reddish. On *Fabaceae* species ›› 21

20*
Stigma purple or brown, very rarely yellow. Not on *Fabaceae* species ›› 22

21
Stigma yellow. Corolla brown, yellowish below, frequently with violet veins, rarely purple or yellowish all over. ›› 4.22 *O. lutea* Baumg.

21*
Stigma yellow at first, later reddish. Corolla purple-brown or blackish, except at base, inside brightly coloured and shiny. Middle lobe of the lower lip of the corolla about twice the size of the side lobes. Mainly in the western Mediterranean region to northern Italy ›› 4.30 *O. variegata* Wallr.

22
Stamens inserted near the base of the corolla-tube. Corolla light brown-violet or reddish, rarely yellow, all over. Mostly on *Galium* and *Asperula* species ›› 4.11 *O. caryophyllacea* Smith

22*
Stamens inserted 3-5 mm above the base of the corolla-tube. Corolla light brown-violet, whitish below. On *Teucrium* species ›› 4.29 *O. teucrii* Holandre

23 (9)
Corolla yellow on the outside, upper lip and veins red or purple, sometimes red or purple all over, inside usually shiny purple and brown to black when dried, rarely wax-coloured all over. Stigma usually yellow and with blood-red or purple margin ›› 24

23*
Corolla not shiny purple on the inside ›› 25

24
Staubblätter am Grund der Kronröhre oder wenig darüber eingefügt. Nelkengeruch » 4.18 *O. gracilis* Smith

24*
Staubblätter 2 bis 4 mm hoch über dem Grund der Kronröhre eingefügt. Geruch unangenehm. Mittellappen der Unterlippe der Blütenkrone meistens doppelt so groß wie die beiden Seitenlappen. Verbreitung vor allem westliches Mittelmeergebiet bis Norditalien » 4.30 *O. variegata* Wallr.

25
Blumenkrone weiß oder bleich, gegen den Saum zu mit blauen oder violetten Nerven, getrocknet weißlich oder hellbraun; Rückenlinie in der Mitte gerade oder schwach gebogen. Narbe hellviolett, hellrot, hellgelb oder weißlich. Selten eingeschleppt. Meist auf kultivierten *Fabaceae*- oder *Pelargonium*-Arten » 4.14 *O. crenata* Forsk.

25*
Blütenfarbe anders. Rückenlinie gleichmäßiger gebogen. Narbe gelb, violett oder rotbraun » 26

26
Narbe gelb. Blumenkrone 10 bis 30 mm lang » 27

26*
Narbe meist violett oder rotbraun. Blumenkrone nicht über 20 mm lang. Meist auf *Trifolium*-Arten und *Eryngium campestre* » 36

27
Staubblätter höchstens 2 mm hoch über dem Grund der Kronröhre eingefügt, unten kahl. Auf *Cytisus*-, *Genista*- und *Ulex*-Arten » 4.26 *O. rapum-genistae* Thuill.

27*
Staubblätter fast immer höher eingefügt, unten behaart » 28

28
Kronröhre unter dem Schlund zusammengezogen, weiß oder gelblich, an der Oberlippe rötlich, oft mit lila Adern. Vor allem auf *Hedera helix* » 4.19 *O. hederae* Duby

28*
Kronröhre nach vorn zu allmählich erweitert; Blumenkrone mäßig gekrümmt, meist aufrecht abstehend » 29

29
Blütenstand zur Gänze dichtblütig oder höchstens im unteren Teil locker. Blumenkrone 12 bis 27 mm lang. Auf *Umbelliferae*- und *Centaurea*-Arten » 30

29*
Blütenstand nur anfangs dichtblütig, bald im unteren Teil oder zur Gänze lockerblütig. Blumenkrone 12 bis 22 mm lang. Auf anderen Wirtspflanzen. Im Berg- und Voralpengebiet » 34

30
Kronröhre über der Einfügungsstelle der Staubblätter allmählich erweitert, zuerst rosenrot, später bleichgelb. Oberlippe nicht oder wenig ausgerandet. Auf *Centaurea*-Arten, vor allem auf *Centaurea scabiosa* » 4.16 *O. elatior* Sutt.

30*
Kronröhre über der Einfügungsstelle der Staubblätter plötzlich bauchig erweitert. Oberlippe ausgerandet bis tief zweilappig » 31

24
Stamens inserted at or near the base of the corolla-tube. Scent of carnation » 4.18 *O. gracilis* Smith

24*
Stamens inserted 2-4 mm above the base of the corolla-tube. Fetid smell. Middle lobe of the lower lip of the corolla about twice the size of the side lobes. Mainly in the western Mediterranean region to northern Italy » 4.30 *O. variegata* Wallr.

25
Corolla white or pale, with blue or violet veins near the margin, whitish or light brown when dried; dorsal line of the corolla straight or slightly curved in the middle. Stigma light violet, light red, light yellow or whitish. Rarely introduced. Mostly on cultivated *Fabaceae* or *Pelargonium* species » 4.14 *O. crenata* Forsk.

25*
Colour of corolla different. Dorsal line of the corolla more evenly curved. Stigma yellow, violet or red-brown » 26

26
Stigma yellow. Corolla 10-30 mm » 27

26*
Stigma usually violet or red-brown. Corolla less than 20 mm. Mostly on *Trifolium* species and on *Eryngium campestre* » 36

27
Stamens inserted less than 2 mm above the base of the corolla-tube, glabrous below. On *Cytisus*, *Genista* and *Ulex* species » 4.26 *O. rapum-genistae* Thuill.

27*
Stamens usually inserted higher, pubescent below » 28

28
Corolla-tube constricted below the throat, white or yellowish, reddish at the upper lip, frequently with violet veins. Mainly on *Hedera helix* » 4.19 *O. hederae* Duby

28*
Corolla-tube widening gradually towards the throat; corolla slightly curved, usually erect to spreading » 29

29
Inflorescence dense, sometimes lax below. Corolla 12-27 mm. On *Umbelliferae* and *Centaurea* species » 30

29*
Inflorescence dense only in the beginning, soon lax below or all over. Corolla 12-22 mm. On other hosts. In mountainous regions and alpine foothills » 34

30
Corolla-tube gradually widening above insertion of stamens, rosy red at first, pale yellow later. Upper lip not or only slightly emarginate. On *Centaurea* species, mainly on *Centaurea scabiosa* » 4.16 *O. elatior* Sutt.

30*
Corolla-tube inflated abruptly above insertion of the stamens. Upper lip emarginate to deeply bilobate » 31

31
Blumenkrone 25 mm lang. Staubblätter oben drüsenhaarig. Stengel sehr kräftig, mindestens 40 cm hoch. Vor allem auf *Laserpitium siler*. Mittel- und Osteuropa (Frankreich, die Schweiz, Liechtenstein, Österreich, Slowenien, Balkan, Pyrenäen) » 4.20 *O. laserpitii-sileris* Reut. ex Jordan

31*
Stengel im oberen Teil deutlich schwächer beschuppt als im unteren- und mittleren Teil. Staubblätter oben fast kahl. Wirt entweder *Seseli*, *Peucedanum* oder *Laserpitium* » 32

32
Blüten größer als 20 mm. Staubblätter mindestens 4 mm hoch über dem Grund der Kronröhre eingefügt. Griffel deutlich drüsenhaarig. Wirt vorwiegend *Peucedanum cervaria* » 4.6 *O. alsatica* Kirschl.

32*
Blüten kleiner als 20 mm. Staubblätter 1 bis 4 mm hoch über dem Grund der Kronröhre eingefügt. Griffel meist kahl » 33

33
Pflanze zierlich. Blütenkrone oft violettrötlich bis rotbraun (rosa) gefärbt. Staubblätter 1 bis 3 mm hoch über dem Grund der Kronröhre eingefügt. Auf *Seseli libanotis* » 4.10 *O. bartlingii* Griseb.

33*
Stengel und Blumenkrone meist reingelb. Kelchhälften vorn meist nicht verwachsen. Staubblätter 2 bis 4 mm hoch über dem Grund der Kronröhre eingefügt. Nur auf der Schwäbischen Alb und im Maingebiet bei Karlstadt in Süddeutschland. Auf *Laserpitium latifolium* » 4.7 *O. alsatica* Kirschl. subsp. *mayeri* (Ssg. et Ronn.) C.A.J. Kreutz

34 (29*)
Oberlippe der Blumenkrone mit am Rande fast kahlen, zuletzt zurückgeschlagenen Zipfeln. Blumenkrone hellgelb oder gelblichweiß, an der Oberlippe rötlich. Staubblätter oben drüsenhaarig. Griffel kahl oder spärlich drüsenhaarig. Auf *Petasites-*, *Tussilago-* und *Adenostyles-*Arten » 4.17 *O. flava* Mart. ex F.W. Schultz

34*
Oberlippe der Blumenkrone mit am Rande drüsenhaarigen, vorgestreckten oder fast abstehenden Zipfeln. Staubblätter oben fast kahl » 35

35
Lappen der Oberlippe zuletzt abstehend. Griffel reichlich drüsenhaarig. Auf *Salvia-*Arten (meist *Salvia glutinosa*) » 4.28 *O. salviae* F.W. Schultz ex Koch

35*
Lappen der Oberlippe gerade vorgestreckt. Griffel meist kahl oder spärlich drüsenhaarig. Auf *Berberis-* und *Rubus-*Arten » 4.21 *O. lucorum* A. Br. ex Koch

36 (26*)
Blüten abstehend bis aufrecht-abstehend, ca. 5 bis 8 mm breit, purpurn überlaufen, drüsenhaarig bis fast kahl. Auf verschiedenen Wirtspflanzen » 37

36*
Pflanze violett gefärbt, bis zu 60 cm hoch. Blütenstand im unteren Teil locker- und im oberen Teil dichtblütig. Narbe rotbraun bis violett. Auf verschiedenen Wirtspflanzen. Verbreitung vor allem in Ost-Anglia und Surrey (England) » 38

31
Corolla 25 mm. Stamens glandular-pubescent above. Stem very stout, at least 40 cm tall. Mainly on *Laserpitium siler*. Central and eastern Europe (France, Switzerland, Liechtenstein, Austria, Slovenia, Balkan countries, Pyrenees) » 4.20 *O. laserpitii-sileris* Reut. ex Jordan

31*
Stem more sparsely scaled above than in the middle and below. Stamens almost glabrous above. Host either *Seseli*, *Peucedanum* or *Laserpitium* » 32

32
Flowers larger than 20 mm. Stamens inserted more than 4 mm above the base of the corolla-tube. Stigma clearly glandular-pubescent. Host usually *Peucedanum cervaria* » 4.6 *O. alsatica* Kirschl.

32*
Flowers smaller than 20 mm. Stamens inserted 1-4 mm above the base of the corolla-tube. Stigma usually glabrous. » 33

33
Plant graceful. Corolla often violet to reddish or reddish-brown (pink). Stamens inserted 1-3 mm above the base of the corolla-tube. On *Seseli libanotis* » 4.10 *O. bartlingii* Griseb.

33*
Stem and corolla usually purely yellow. Calix-segments not fused at front. Stamens inserted 2-4 mm above the base of the corolla-tube. Only in southern Germany (on the Schwäbische Alb and in the Main area near Karlstadt). On *Laserpitium latifolium* » 4.7 *O. alsatica* Kirschl. subsp. *mayeri* (Ssg. et Ronn.) C.A.J. Kreutz

34 (29*)
Margin of the upper lip of the corolla with nearly glabrous, ultimately recurved tips. Corolla bright yellow or yellowish-white, reddish at the upper lip. Stamens glandular-pubescent above. Stigma glabrous or sparsely glandular-pubescent. On *Petasites*, *Tussilago* and *Adenostyles* species » 4.17 *O. flava* Mart. ex F.W. Schultz

34*
Upper lip of the corolla with porrect or almost erecto-patent edges, margin glandular-pubescent. Stamens almost glabrous above » 35

35
Lobes of the upper lip spreading at maturity. Stigma densely glandular-pubescent. On *Salvia* species (mostly *Salvia glutinosa*) » 4.28 *O. salviae* F.W. Schultz ex Koch

35*
Lobes of the upper lip porrect. Stigma mostly glabrous or sparsely glandular-pubescent. On *Berberis* and *Rubus* species » 4.21 *O. lucorum* A. Br. ex Koch

36 (26*)
Flowers spreading to erect-spreading, approximately 5-8 mm wide, tinged with purple, glandular-pubescent to almost glabrous. On various hosts » 37

36*
Plant violet, up to 60 cm tall. Inflorescence lax below and dense above. Stigma red-brown to violet. On various hosts. Mainly in East-Anglia and Surrey (England) » 38

37
Mittellappen der Unterlippe der Blütenkrone nierenförmig und größer als die beiden Seitenlappen. Blüten etwa 6 bis 10 mm groß, unterste Blüten meistens gestielt. Die zwei halbkugeligen Lappen der Narbe teilweise miteinder verbunden. Auf *Daucus carota*, selten auf *Plantago coronopus* und *Ononis repens*. Nur in Südengland, den Kanal-Inseln und in Westfrankreich » 4.23 *O. minor* Sm. var. *maritima* (Pugsley) Rumsey & Jury

37*
Blumenkrone 10 bis 17, meist 15 mm lang, gegen den Schlund wenig erweitert, Lappen der Oberlippe vorgestreckt oder abstehend, Staubblätter 2 bis 3 mm hoch über dem Grund der Kronröhre eingefügt. Verbreitung fast ganz Europa. Vorwiegend auf *Trifolium*-Arten » 4.23 *O. minor* Sm. s.l.

38
Blüten bleich violett gefärbt und spärlich mit Drüsenharen besetzt. Blütenkrone 12 bis 18 mm lang und 3,5 bis 5 mm breit. Staubblätter an der Basis dicht behaart. Auf *Asteraceae*-Arten, hauptsächlich auf *Crepis virens*, *Hypochoeris radicata*, *Tripleurospermum inodorum*, *Carduus nutans* und *Senecio greyii* » 4.23 *O. minor* Sm. var. *compositarum* Pugsley

38*
Blumenkrone aus gekrümmter Basis gerade oder nach vorne gebogen, später von der Einfügungsstelle der Staubblätter an knieförmig oder abwärts gebogen, weiß, oben an den Nerven violett. Oberlippe tief zweispaltig mit abstehenden, später zurückgeschlagenen Lappen. Nur auf *Eryngium campestre*. Nur in Mittel- und Südeuropa » 4.8 *O. amethystea* Thuill.

Dieser Bestimmungsschlüssel wurde zum Teil übernommen aus Hegi G., *Illustrierte Flora von Mitteleuropa*, Band 6, Seite 475-477, 1974.
Außerdem wurde der Bestimmungsschlüssel ergänzt mit *Orobanche bartlingii*, *O. cumana*, *O. pallidiflora*, *O. purpurea* var. *bohemica*, *O. minor* var. *maritima*, *O. minor* var. *compositarum* und *O. alsatica* subsp. *mayeri*.
Die Sippen *O. minor* var. *maritima* und *O. minor* var. *compositarum*, sowie *O. purpurea* var. *bohemica* wurden in den Bestimmungsschlüssel aufgenommen, weil sie in den meisten englischen, beziehungsweise in den tschechischen und slowakischen Floren als eigene Arten aufgeführt werden.

37
Middle lobe of the lower lip of the corolla reniform and larger than other lobes. Flowers approximately 6-10 mm, lower flowers usually pediculate. Hemispherical lobes of the stigma partly fused. On *Daucus carota*, rarely on *Plantago coronopus* and *Ononis repens*. Only in southern England, on the Channel Islands and in western France » 4.23 *O. minor* Sm. var. *maritima* (Pugsley) Rumsey & Jury

37*
Corolla 10-17 mm, mostly 15 mm, slightly widening towards the throat, lobes of the upper lip spreading or erecto-patent, stamens inserted 2-3 mm above the base of the corolla-tube. Throughout most parts of Europe. Mainly on *Trifolium* species » 4.23 *O. minor* Sm. s.l.

38
Flowers pale violet and sparsely glandular-pubescent. Corolla 12-18 mm and 3.5-5 mm wide. Stamens densely pubescent at the base. On *Asteraceae* species, especially on *Crepis virens*, *Hypochoeris radicata*, *Tripleurospermum inodorum*, *Carduus nutans* and *Senecio greyii* » 4.23 *O. minor* Sm. var. *compositarum* Pugsley

38*
Corolla straight or curved to the front from curved base, later geniculate or bent downwards from the insertion of the stamens on, white, violet at the veins above. Upper lip deeply bifid with spreading lobes, later recurved. Only on *Eryngium campestre*. Only in central and southern Europe » 4.8 *O. amethystea* Thuill.

This key has partly been copied from Hegi G., *Illustrierte Flora von Mitteleuropa*, Band 6, 475-477, 1974.
Furthermore, the following species have been added: *Orobanche bartlingii*, *O. cumana*, *O. pallidiflora*, *O. purpurea* var. *bohemica*, *O. minor* var. *maritima*, *O. minor* var. *compositarum* and *O. alsatica* subsp. *mayeri*.
The taxa *O. minor* var. *maritima* and *O. minor* var. *compositarum* have been incorporated here, because they are listed as species in most of the English floras. *O. purpurea* var. *bohemica* has been included because it is regarded as a species in the Czech and Slovak floras.

4.1 SEKTION/SECTION TRIONYCHON WALLROTH 1822

OROBANCHE ARENARIA — BORKHAUSEN 1794

Orobanche laevis L.; *Kopsia arenaria* (Borkh.) Dumortier; *Phelipaea arenaria* (Borkh.) Walpers

Sand-Sommerwurz / Sand Broomrape

- **ARTBESCHREIBUNG**

Pflanzen meistens kräftig (selten schlank) und niedrig, etwa 15 bis 55 cm hoch. Der Stengel ist kräftig oder schlank, aufrecht (selten ästig), (hell)gelb, gelblich-weiß, gelblich oder blaßlila gefärbt, mit kurzen Drüsenhaaren besetzt, unten dicht und reichlich (dachig), oben spärlicher und lockerer beschuppt. Die Schuppen sind breitlanzettlich, aufrecht bis abstehend, dunkelbraun gefärbt, spärlich drüsenhaarig oder kahl. Der Blütenstand ist meist dicht- und reichblütig oder arm- und lockerblütig, zylindrisch. Die zwei Vorblätter sind lanzettlich, gelblich(weiß) gefärbt, etwa ein Drittel bis halb so lang wie die Blütenkrone (etwa gleich lang wie die Kelchzähne) mit weißen Drüsenhaaren. Das Tragblatt ist lanzettlich (am Grunde eiförmig), etwa halb so lang wie die Blütenkrone, im oberen Teil manchmal etwas abwärts gekrümmt, gelblich (an der Spitze dunkelbraun) gefärbt und reichlich mit weißen Drüsenhaaren besetzt. Kelch röhrig-glockig, aus vier- bis fünf Zähnen bestehend, reichlich drüsenhaarig, der hintere Kelchzahn sehr klein, die anderen etwa ein Drittel bis halb so lang wie die Blütenkrone (Kelchzähne ungefähr so lang wie die Kronröhre) und (hell)gelblich gefärbt. Die Blüten sind mittelgroß bis groß, erst aufrecht, später abstehend. Die Blütenkrone ist 25 bis 35 mm lang, röhrig, über dem unteren etwas blasig erweiterten Drittel verengt und in der mittleren und oberen Hälfte trichterig erweitert, mit hellen Drüsenharen besetzt, bläulich, lila oder blauviolett (am Grunde oft heller) gefärbt mit dunkelvioletten Nerven. Die Rückenlinie der Blütenkrone ist vom Grund an fast gleichmäßig gebogen, in der Mitte fast gerade und an der Spitze manchmal ein wenig aufgerichtet. Die Oberlippe der Blütenkrone ist zweilappig, behaart und mit aufgerichteten Lappen. Die Unterlippe der Blütenkrone ist herabgeschlagen, mit drei fast gleichgroßen, kreisrundlichen, gerundeten mit weißhaarigen Falten versehenen Lappen. Die Staubblätter sind 5 bis 7 mm hoch über dem Grund der Kronröhre eingefügt. Die Staubfäden sind an der Basis fast kahl oder unten bis zu den Staubbeuteln spärlich drüsenhaarig. Die Staubbeutel sind lang wollig behaart. Der Griffel ist reichlich drüsenhaarig. Die Narbe besteht aus zwei Lappen und ist weiß gefärbt. 2n = 24.

- **SPECIES DESCRIPTION**

The plant is usually robust (rarely slender) and small, approximately 15-55 cm tall. The stem is robust or slender, erect (rarely branched), (bright) yellow, yellowish-white, yellowish or pale violet, with short glandular hairs, with numerous (imbricate) scale leaves below, sparsely and laxly scaled above. The scale leaves are broadly lanceolate, spreading to erecto-patent, dark brown and sparsely glandular or glabrous. The inflorescence is cylindrical, with numerous flowers in a dense spike or with fewer ones in a lax spike. The two bracteoles are lanceolate, yellowish (white), about a third to one half of the length of the corolla (about the size of the calyx-segments), with white glandular hairs. The bract is lanceolate (oval at the base), approximately half the size of the corolla, sometimes slightly deflexed above, yellowish (dark brown at the tip) with many white glandular hairs. The calyx is tubular-campanulate, 4- or 5-dentate and densely glandular-pubescent. The rear calyx-segment is very small, the others are a third to one half of the size of the corolla (calyx-segments are about as long as the corolla-tube); all are (bright) yellow. The flowers are of medium to large size, erect at first, erecto-patent later. The corolla is 25-35 mm long, tubular, slightly inflated in the lower third part, then constricted, and funnel-shaped in the middle and upper part, pubescent with light glandular hairs, bluish, violet or blue-violet (frequently brighter at the base) with dark violet veins. The dorsal line of the corolla is nearly evenly curved from the base onwards, almost straight in the middle and sometimes bent upwards at the tip. The upper lip of the corolla is bilobate, pubescent with erect lobes. The lower lip of the corolla is deflexed and has three circular, rounded, white, plicate lobes of almost equal size, with white glandular hairs in the folds. The stamens are inserted 5-7 mm above the base of the corolla-tube. The filaments are almost glabrous at the base, or sparsely glandular below, up to the anthers. The anthers are woolly, with long hairs. The style has many glandular hairs. The stigma consists of two lobes and is white. 2n = 24.

- **BLÜTEZEIT**

Mitte Mai bis Ende Juni in wärmeren Gegenden, sonst im Juli.

- **FLOWERING TIME**

Mid-May to end of June in temperate regions, July elsewhere.

- **STANDORT**

Auf Trocken-, (Halbtrocken-), Sand-, Steppen- und Xerothermrasen, an lichten Gebüschsäumen, in grasigen Hängen und in Felsfluren an warmen, sonnigen Standorten auf trockenen, nährstoffarmen Sand- und Lößböden.

- **HABITAT**

In arid and semi-arid grassland, sandy and xerothermic grassland and steppes, in open thickets, on grassy slopes and rocky ground, in sunny, warm places on dry, nutrient-poor sandy soil and loess.

- **WIRT**

Schmarotzt auf *Artemisia campestris*, selten auf *A. vulgaris*.

- **HOST**

Parasitic on *Artemisia campestris*, rarely on *A. vulgaris*.

- **GESAMTVERBREITUNG**

Mittel-, Süd- und Osteuropa; von Portugal und Spanien über Frankreich, Norditalien durch Mitteleuropa bis zum Kaukasus, Kleinasien und Iran, weiter bis Zentralasien. Nördlich vereinzelt bis Nordostfrankreich, Nordostdeutschland bis West- und Südpolen. Auch in Nordafrika (Marokko, Algerien und Tunesien). Nicht auf den Mittelmeerinseln und in Griechenland. Angaben von *Orobanche arenaria* aus Estland beruhen wahrscheinlich auf eine Verwechslung mit *O. purpurea*. *Orobanche arenaria* besitzt ihre Standorte hauptsächlich in den Steppen Südrußlands; die Fundortdichte ist aber in der ehemalige UdSSR geringer als in den Trockenrasen Mitteleuropas ostwärts bis Ungarn und Rumänien (Beck-Mannagetta, 1930). In den anderen Teilen Europas ist sie sehr selten und tritt zerstreut auf.

- **DISTRIBUTION**

Central, southern and eastern Europe; from Portugal and Spain, through France, northern Italy, central Europe to the Caucasus, Asia Minor and Iran, further to central Asia. To the north sporadically up to north-eastern France, north-eastern Germany and to western and southern Poland. Also in northern Africa (Morocco, Algeria and Tunisia). Not on Mediterranean islands and not in Greece. *Orobanche arenaria* finds in Estonia were probably *O. purpurea* erroneously identified as *O. arenaria*.
O. arenaria mainly grows in the steppes of southern Russia, but the density of locations in the former USSR is lower than in arid grassland in central Europe eastward to Hungary and Rumania (Beck-Mannagetta, 1930). The species is very rare and sporadic in other parts of Europe.

- **BEMERKUNGEN**

Orobanche arenaria ist schwierig von *O. purpurea* zu unterscheiden. *O. arenaria* hat vor allem viel größere Blüten, wollig behaarte Staubbeutel und einen anderen Wirt; außerdem sind ihre Blüten im oberen Teil der Blumenkrone weit trichterig erweitert.

- **COMMENTS**

It is difficult to distinguish *Orobanche arenaria* from *O. purpurea*. *O. arenaria* has much larger flowers with woolly anthers and is parasitic on a different host; also, its upper corolla is widely funnel-shaped in the upper part.

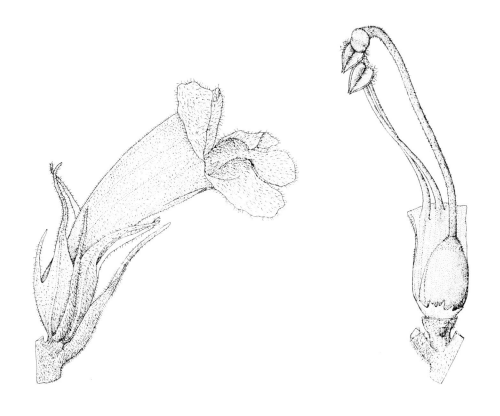
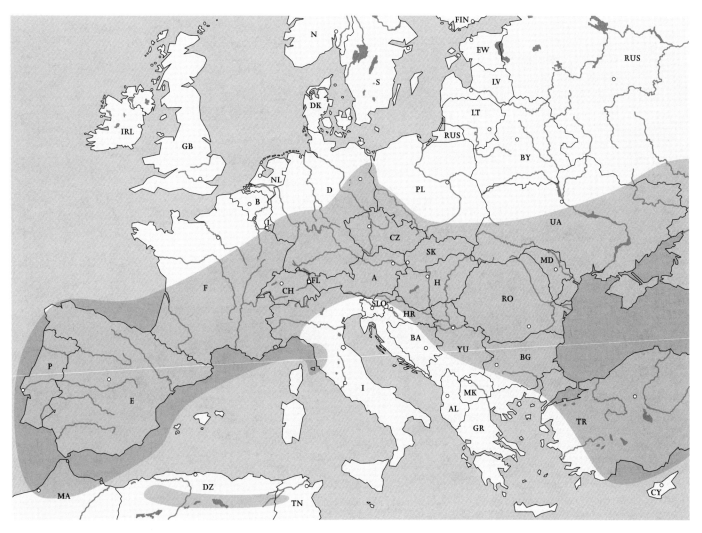

OROBANCHE ARENARIA

Waldböckelheim, Nordpfälzer Bergland (D), 1-7-1989

OROBANCHE ARENARIA

mit Wirtspflanze / with host (*Artemisia campestris*), Waldböckelheim, Nordpfälzer Bergland (D), 18-6-1994

Waldböckelheim, Nordpfälzer Bergland (D), 18-6-1994

Waldböckelheim, Nordpfälzer Bergland (D), 18-6-1994

Waldböckelheim, Nordpfälzer Bergland (D), 1-7-1989

4.2

OROBANCHE CAESIA — REICHENBACH 1829

Orobanche peisonis G. Beck; *O. lanuginosa* Beck; *Phelipaea caesia* (Rchb.) Reuter

Weißwollige (Blaugraue) Sommerwurz

- **Artbeschreibung**

Pflanzen meistens kräftig und niedrig, etwa 10 bis 30 cm hoch. Der Stengel ist kräftig, aufrecht, gelb, gelblich, orangegelb, violett oder bläulichbraun gefärbt, drüsenhaarig bis weißwollig (auch im oberen Teil), unten reichlich, oben spärlich beschuppt. Die Schuppen sind breitlanzettlich (an der Basis eiförmig), aufrecht, am Grunde gelblich und im mittleren und oberen Teil dunkelbraun gefärbt, drüsenhaarig. Der Blütenstand ist meist dicht- und reichblütig, zylindrisch oder kurz. Die zwei Vorblätter sind lanzettlich, gelblich gefärbt, etwa ein Drittel bis halb so lang wie die Blütenkrone (ungefähr so lang wie die Kelchzähne) mit weißen Drüsenhaaren (weißwollig). Das Tragblatt ist lanzettlich (am Grunde eiförmig), im oberen Teil abwärts gekrümmt, gelblich (an der Spitze dunkelbraun) gefärbt, etwa halb so lang wie die Blütenkrone und reichlich mit weißen Drüsenhaaren (weißwollig) besetzt. Kelch röhrig-glockig, aus vier bis fünf Zähnen bestehend, reichlich drüsenhaarig und weißwollig, meist ein Drittel bis halb so lang wie die Blütenkrone (Kelchzähne ungefähr so lang wie die Kronröhre) und gelblich gefärbt. Die Blüten sind mittelgroß, erst aufrecht bis abstehend, später fast waagrecht abstehend oder stark vorwärts gekrümmt. Die Blütenkrone ist 15 bis 25 mm lang, röhrig, im unteren Teil etwas eingeschnürt und in der oberen Hälfte trichterig erweitert, reichlich mit Drüsenhaaren besetzt, bläulich, lila oder blauviolett gefärbt mit dunkelvioletten Nerven. Die Rückenlinie der Blütenkrone ist vom Grund an fast gleichmäßig gebogen und an der Spitze manchmal ein wenig aufgerichtet. Die Oberlippe der Blütenkrone ist tief zweilappig mit vorgestreckten oder schwach aufgerichteten Lappen. Die Unterlippe der Blütenkrone ist herabgeschlagen, mit drei fast gleichgroßen, länglichen, gerundeten, gezähnelten Lappen. Die Staubblätter sind über den verengten Teil der Blütenkrone 6 bis 8 mm hoch eingefügt. Die Staubfäden sind fast kahl oder an der Basis spärlich mit Drüsenhaaren besetzt. Die Staubbeutel sind spärlich mit Haaren besetzt oder fast kahl. Der Griffel ist drüsenhaarig. Die Narbe besteht aus zwei kugeligen Lappen und ist weiß gefärbt. $2n = 24$.

- **Blütezeit**

Mitte Mai bis Juli.

- **Standort**

Auf Trocken-, (Halbtrocken-), Steppen-, und Xerothermrasen, an lichten Gebüschsäumen, an grasigen Hängen, an warmen, sonnigen Standorten auf basenreichen, sandigen Böden.

- **Wirt**

Schmarotzt auf *Artemisia pontica* und *A. austriaca*.

- **Gesamtverbreitung**

Orobanche caesia ist eine typische Steppenpflanze. Ihr Verbreitungsgebiet reicht von Osteuropa, Österreich (Niederösterreich und Burgenland), der Tschechischen Republik (südöstlicher Teil), die Slowakei, Ungarn, Rumänien und Moldawien) über die Kaukasusländer bis Zentralasien.

Die Art ist vor allem im östlichen Teil von Mitteleuropa sehr selten und dort nur noch auf wenige Standorte beschränkt.

- **Bemerkungen**

Orobanche caesia unterscheidet sich von den anderen Arten der Sektion *Trionychon* (*O. ramosa*, *O. arenaria* und *O. purpurea*) durch ihre weißwollige Behaarung und ihre auffallenden, etwas aufgeblasenen Blüten. Außerdem ist ihr Areal in Europa sehr begrenzt.

Blue-grey Broomrape

- **Species description**

The plant is usually robust and small, approximately 10-30 cm tall. The stem is robust, erect, yellow, yellowish, orange-yellow, violet or bluish-brown, with glandular hairs or white woolly (also in the upper parts), with numerous scale leaves below, sparsely scaled above. The scale leaves are broadly lanceolate (oval at the base), erect, yellowish at the base and dark brown in the middle part and above; glandular. The inflorescence usually has numerous flowers in a dense, cylindrical or short spike. The two bracteoles are lanceolate, yellowish, about a third to one half of the length of the corolla (about the size of the calyx-segments) with white glandular hairs. The bract is lanceolate (oval at the base), deflexed in the upper part, yellowish (dark brown at the tip), approximately half the length of the corolla and richly woolly with white glandular hairs. The calyx is tubular-campanulate, 4- or 5-dentate, densely glandular-pubescent and woolly with white hairs, usually a third to one half of the length of the corolla (length of calyx-segments equal to corolla-tube), yellowish. The flowers are of medium size, erect or erecto-patent at first, almost horizontal or clearly flexed forward later. The corolla is 15-25 mm long, tubular, slightly contracted below and funnel-shaped in the upper half, richly pubescent with glandular hairs, bluish, violet or blue-violet with dark violet veins. The dorsal line of the corolla is almost evenly curved from the base onwards and sometimes bent upwards at the tip. The upper lip of the corolla is deeply bilobate with porrect or suberect lobes. The lower lip of the corolla is deflexed and has three elongated, oval, crenate lobes of almost equal size. The stamens are inserted 6-8 mm above the contracted part of the corolla-tube. The filaments are almost glabrous or sparsely glandular-pubescent at the base. The anthers are sparsely pubescent or almost glabrous. The style is glandular-pubescent. The stigma consists of two spherical lobes and is white. $2n = 24$.

- **Flowering time**

Mid-May to July.

- **Habitat**

In arid and semi-arid grassland, steppes and xerothermic grassland, in open thickets and on grassy slopes in sunny, warm places on alkaline, sandy soil.

- **Host**

Parasitic on *Artemisia pontica* and *A. austriaca*.

- **Distribution**

Orobanche caesia typically occurs on grasslands (steppe). Its range covers eastern Europe, Austria (Niederösterreich und Burgenland), the Czech Republic (south-eastern part), Slovakia, Hungary, Rumania and Moldavia), the Caucasus countries and central Asia.

The species is very rare in eastern central Europe, where it is limited to a few locations.

- **Comments**

Orobanche caesia distinguishes itself from other species of the section *Trionychon* (*O. ramosa*, *O. arenaria* and *O. purpurea*) by its white, woolly pubescence and its conspicuous, slightly inflated flowers. Furthermore, its European range is very limited.

OROBANCHE CAESIA

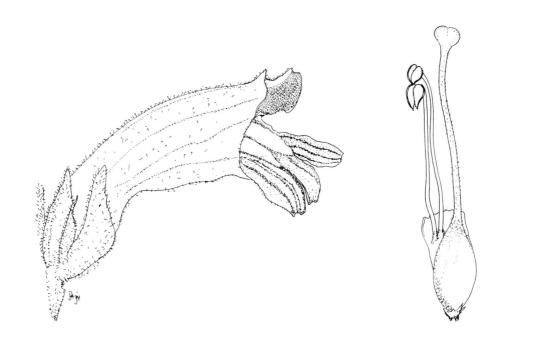

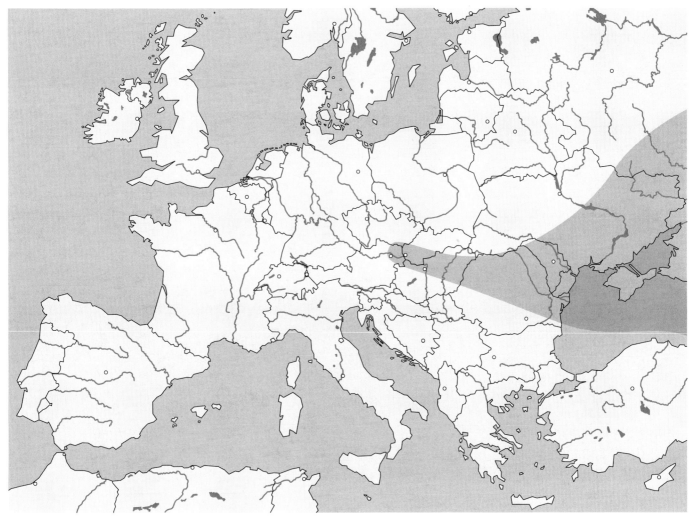

mit Wirtspflanze / with host (*Artemisia austriaca*), Jois, Burgenland (A), 28-5-1994

mit Wirtspflanze / with host (*Artemisia austriaca*), Jois, Burgenland (A), 28-5-1994

Jois, Burgenland (A), 28-5-1994

Jois, Burgenland (A), 28-5-1994

Jois, Burgenland (A), 28-5-1994

4.3

OROBANCHE PURPUREA — JACQUIN 1762
Orobanche caerulea Villars; *Philipaea purpurea* (Jacq.) Ascherson; *Philipaea caerulea* (Vill.) C.A. Meyer

Purpur- (Violette) Sommerwurz

- **ARTBESCHREIBUNG**

Pflanzen meistens schlank, selten kräftig, etwa 15 bis 65 cm hoch. Der Stengel ist meist schlank, aufrecht, blaßlila, violett oder blaugrau gefärbt (oft violett überlaufen), mit kurzen hellen Drüsenhaaren besetzt, unten spärlich, oben sehr locker beschuppt. Die Schuppen sind eiförmig-lanzettlich, aufrecht, blaugrau oder blaßlila gefärbt, reichlich drüsenhaarig. Der Blütenstand ist meist locker- und reichblütig (anfangs dicht und zylindrisch), oft sehr lang und gestreckt, wobei die Blüten meist weit unten am Stengel stehen, selten finden sich auch Exemplare mit zylindrischem Blütenstand, die arm- und lockerblütig sind. Die zwei Vorblätter sind lanzettlich, gelblichbraun oder blaugrau (violett) gefärbt, etwa ein Drittel so lang wie die Blütenkrone (kürzer als die Kelchzähne) mit weißen Drüsenhaaren. Das Tragblatt ist lanzettlich, etwa ein Drittel so lang wie die Blütenkrone, nicht abwärts gekrümmt, blaugrau (violett) gefärbt und reichlich drüsenhaarig. Kelch röhrig-glockig, aus vier bis fünf Zähnen bestehend, reichlich drüsenhaarig, der fünfte Kelchzahn sehr klein (meist verkümmert), die anderen etwa ein Drittel so lang wie die Blütenkrone (Kelchzähne meist kürzer als die Kronröhre) und gelblichbraun oder blaugrau (violett) gefärbt. Die Blüten sind mittelgroß, erst aufrecht, später abstehend. Die Blütenkrone ist 18 bis 25 mm lang, röhrig, über dem unteren etwas aufgeblasenen Drittel schwach eingeschnürt oder verengt und darüber allmählich trichterig erweitert, mit kurzen, hellen Drüsenhaaren besetzt, bläulich, blaßlila oder blauviolett (an der Basis gelblich oder gelblichweiß) gefärbt mit dunkelvioletten Nerven. Die Rückenlinie der Blütenkrone ist vom Grund an fast gleichmäßig gebogen, im Bereich der Oberlippe schwach herabgebogen und an der Spitze manchmal ein wenig aufgerichtet. Die Oberlippe der Blütenkrone ist zweilappig, violett geädert, behaart und mit aufgerichteten Lappen. Die Unterlippe der Blütenkrone ist herabgeschlagen und besitzt drei fast gleichgroße, gezähnelte, spitze Lappen, die violett geädert und behaart sind. Die Staubblätter sind 6 bis 8 mm hoch über dem Grund der Kronröhre eingefügt. Die Staubfäden sind an der Basis kahl oder spärlich drüsenhaarig und bis zu den Staubbeuteln meist spärlich drüsenhaarig. Die Staubbeutel sind kahl oder nur an der Spitze spärlich und kurz behaart. Der Griffel ist reichlich drüsenhaarig. Die Narbe besteht aus zwei Lappen und ist weiß oder lila gefärbt. 2n = 24.

- **BLÜTEZEIT**

Anfang Juni bis Ende Juli.

- **STANDORT**

Auf Trocken- und Halbtrockenrasen, in Felsfluren, in Straßengräben, in ruderal beeinflußten Halbtrockenrasen, auf sandigen Stellen (Dünen) und in trockenen Fettwiesen, an warmen, sonnigen Standorten auf lockeren und nährstoffreichen Lehmböden.

- **WIRT**

Schmarotzt vor allem auf *Achillea*-Arten (*A. millefolium* und *A. collina*), aber auch auf *Artemisia vulgaris*, selten auf *Cirsium acaule* und *Lamiaceae*-Arten.

- **GESAMTVERBREITUNG**

Mittel- und Südeuropa; nördlich bis England, die Niederlande, Dänemark, Südschweden (Skåne, Öland), Nordpolen, die baltischen Staaten (Südestland) und Mittelrußland; südlich bis Portugal, Südspanien, Korsika, Süditalien (Sizilien), Griechenland (Kreta) und Kleinasien bis Transkaukasien und Ostindien. Auch auf den Kanarischen Inseln, in Marokko und Nordamerika.
Die Art ist ziemlich verbreitet und in den meisten Teilen Europas selten.

- **BEMERKUNGEN**

Von *Orobanche purpurea* wurde die Varietät *bohemica* (Celak.) Beck 1890 beschrieben. Pflanzen dieser Varietät sind sehr kräftig (bis zur 70 cm hoch) und sind in allen Teilen stärker gefärbt (dunkelviolett). Ihre Ähren sind ziemlich lang, viel- und dichtblütig. Die Blütenkrone ist 20 bis 25 mm lang. Der Stengel ist reichlicher beschuppt und die Staubbeutel sind kahl. Sie schmarotzt nur auf *Artemisia campestris*. Ihr Verbreitungsgebiet liegt vor allem in Osteuropa (unter anderem Niederösterreich und der Tschechischen Republik), sie kommt aber auch zerstreut in Ostdeutschland, Norditalien (Südtirol) und in der Schweiz vor.
Orobanche purpurea ist schwierig von *O. arenaria* zu unterscheiden, aber *O. purpurea* hat vor allem kleinere Blüten als *O. arenaria*, außerdem sind ihre Staubbeutel nicht wollig behaart, und sie schmarotzt vor allem auf *Achillea*-Arten.

Yarrow (Purple) Broomrape

- **SPECIES DESCRIPTION**

The plant is usually slender, rarely stout, approximately 15-65 cm tall. The stem is usually slender, erect, pale violet, violet or blue-grey (frequently tinged with violet), with short, light glandular hairs, sparsely scaled below, very laxly scaled above. The scale leaves are oval-lanceolate, erect, blue-grey or pale violet, densely glandular-pubescent. The inflorescence is usually lax, with numerous flowers (dense and cylindrical at first), its spike frequently long and elongated, with flowers low on the stem. Rarely, plants are found with a lax, cylindrical inflorescence with few flowers. The two bracteoles are lanceolate, yellowish-brown or blue-grey (violet), about a third of the length of the corolla (shorter than the calyx-segments) with white glandular hairs. The bract is lanceolate, approximately a third of the size of the corolla, not deflexed, blue-grey (violet) and richly glandular-pubescent. The calyx is tubular-campanulate, 4- or 5-dentate and densely glandular-pubescent. The fifth calyx-segment is very small (mostly rudimentary), the others are a third of the size of the corolla (calyx-segments are usually shorter than the corolla-tube), all yellowish-brown or blue-grey (violet). The flowers are of medium size, erect at first, later spreading. The corolla is 18-25 mm long, tubular, slightly inflated in the lower third, constricted just above and gradually opening out like funnel, pubescent with short, light glandular hairs, bluish, pale violet or blue-violet (yellowish or yellowish-white at the base) with dark violet veins. The dorsal line of the corolla is almost evenly curved from the base onwards, slightly deflexed near the upper lip and sometimes bent upwards at the tip. The upper lip of the corolla is bilobate with violet veins, pubescent with erect lobes. The lower lip of the corolla is deflexed and has three crenate, acute, pubescent lobes of almost equal size, all having violet veins. The stamens are inserted 6-8 mm above the base of the corolla-tube. The filaments are almost glabrous or sparsely glandular at the base and usually sparsely glandular up to the anthers. The anthers are glabrous or sparsely pubescent (with short hairs) only near the top. The style is richly glandular-pubescent. The stigma consists of two lobes and is white or violet. 2n = 24.

- **FLOWERING TIME**

Beginning of June to end of July.

- **HABITAT**

In arid and semi-arid grassland, on rocky ground, in roadside ditches, in ruderal semi-arid grassland, in sandy places (dunes) and in dry, nutrient-rich meadows, in sunny, warm places on light, nutrient-rich, loamy soil.

- **HOST**

Mainly parasitic on *Achillea* species (*A. millefolium* and *A. collina*), but also on *Artemisia vulgaris*, rarely on *Cirsium acaule* and *Lamiaceae* species.

- **DISTRIBUTION**

Central and southern Europe; northwards to England, the Netherlands, Denmark, southern Sweden (Skåne, Öland), northern Poland, the baltic states (southern Estonia) and central Russia; southwards to Portugal, southern Spain, Corsica, southern Italy (Sicily), Greece (Crete) and Asia Minor to the Transcaucasian area and eastern India. Also on the Canary Islands, in Morocco and in North America.
The species has a wide distribution and is rare in most parts of Europe.

- **COMMENTS**

Of *Orobanche purpurea* a variety *bohemica* (Celak.) Beck 1890 has been described. This plant is very stout (up to 70 cm tall) and more intensely coloured in all parts (dark violet). Its spike is quite long, dense, with many flowers. The corolla is 20-25 mm long. The stem is more richly scaled and the anthers are glabrous. It is parasitic exclusively on *Artemisia campestris*. Its range is mainly in eastern Europe (Niederösterreich (Austria) and the Czech Republic), but it is also found sporadically in eastern Germany, northern Italy (Alto Adige) and Switzerland.
It is not easy to distinguish *Orobanche purpurea* from *O. arenaria*. The main distinguishing features are that *O. purpurea* has smaller flowers than *O. arenaria*, its anthers are not woolly and it is parasitic mainly on *Achillea* species.

OROBANCHE PURPUREA

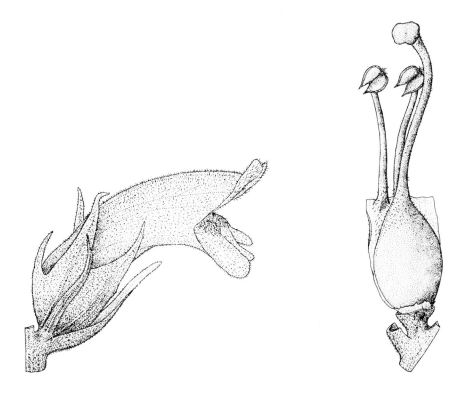

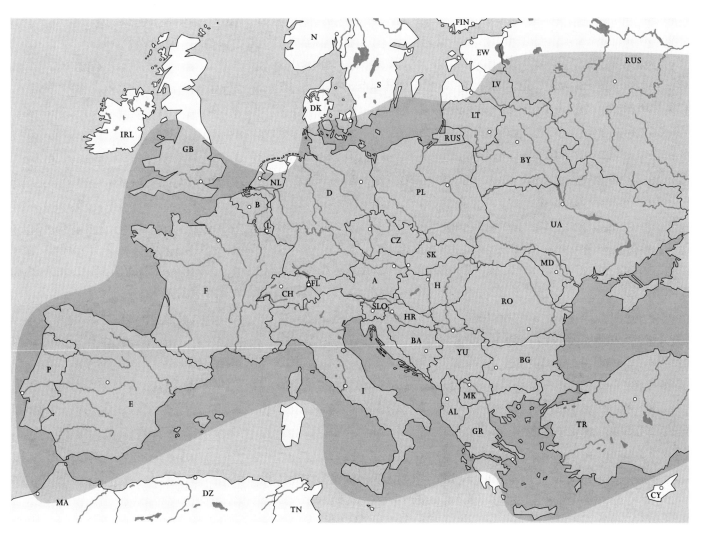

OROBANCHE PURPUREA

Noordwijk aan Zee, Zuid-Holland (NL), 20-6-1988

OROBANCHE PURPUREA

Bomal, Luxembourg (B), 13-6-1988

Noordwijk aan Zee, Zuid-Holland (NL), 20-6-1988

Wöllstein, Hunsrück (D), 1-7-1989

Blankenheim, Eifel (D), 31-7-1984

4.4

OROBANCHE RAMOSA — LINNAEUS 1753

Phelipaea ramosa (L.) C.A. Meyer; *Kopsia ramosa* (L.) Dumortier

Ästige Sommerwurz

- **ARTBESCHREIBUNG**

Pflanzen meistens schlank, etwa 8 bis 45 cm hoch. Der Stengel ist schlank, aufrecht und meistens ästig, (hell)gelb, gelblich, braunviolett, bläulich oder blaßlila gefärbt, reichlich drüsenhaarig, spärlich und locker beschuppt. Die Schuppen sind lanzettlich (an der Basis eiförmig-lanzettlich), aufrecht bis abstehend, dunkelbraun gefärbt, spärlich drüsenhaarig oder kahl. Der Blütenstand ist meist (sehr) locker- und reichblütig (Blütenähre meist sehr verlängert) oder arm- und relativ dichtblütig, zylindrisch. Die zwei Vorblätter sind (breit)lanzettlich, violett bis blaßlila oder gelblich gefärbt, etwa ein Drittel so lang wie die Blütenkrone (gleich lang oder etwas kürzer als die Kelchzähne) mit weißen Drüsenhaaren. Das Tragblatt ist lanzettlich (am Grunde eiförmig), etwa ein Drittel so lang wie die Blütenkrone (etwa so lang wie die Kelchzähne), nicht abwärts gekrümmt, gelblichbraun oder braun gefärbt und reichlich mit weißen Drüsenhaaren besetzt. Kelch röhrig-glockig, meistens aus vier Zähnen bestehend, reichlich drüsenhaarig, etwa ein Drittel so lang wie die Blütenkrone (Kelchzähne kürzer als die Kronröhre) und violett bis blaßlila oder (hell)gelblich gefärbt. Die Blüten sind klein, erst aufrecht, später abstehend und nach vorne gebogen. Die Blütenkrone ist 10 bis 15 mm lang, röhrig, über dem Fruchtknoten etwas verengt und darüber allmählich erweitert, auffällig mit hellvioletten Drüsenhaaren besetzt, bläulich, lila oder blauviolett (vor allem zum Saum hin), selten gelblich oder ganz weiß gefärbt mit dunkelvioletten Nerven. Die Rückenlinie der Blütenkrone ist vom Grund an fast gleichmäßig gebogen, im Bereich der Oberlippe schwach herabgebogen und an der Spitze manchmal ein wenig aufgerichtet. Die Oberlippe der Blütenkrone ist zweilappig, behaart und mit vorgestreckten Lappen. Die Unterlippe der Blütenkrone ist herabgeschlagen, mit drei fast gleichgroßen, gerundeten mit weißfarbigen Falten versehenen Lappen, diese reichlich drüsenhaarig. Die Staubblätter sind 2 bis 4 mm hoch über dem Grund der Kronröhre eingefügt. Die Staubfäden sind an der Basis kahl oder spärlich drüsenhaarig. Die Staubbeutel sind kahl oder mit vereinzelten Haaren besetzt. Der Griffel ist spärlich drüsenhaarig. Die Narbe besteht aus mehreren kugeligen Lappen und ist weiß oder hellbläulich, selten gelblich gefärbt. $2n = 24$.

- **BLÜTEZEIT**

Mitte Juni bis September, im mediterranen Bereich schon ab Ende März.

- **STANDORT**

Die meisten Standorte befinden sich auf sandigen Äckern, besonders in Hackkulturen, an Straßenrändern, in Unkrautgesellschaften, in ruderal beeinflußten Wiesen auf nährstoff- und basenreichen (oft kalkhaltigen), sandigen oder lehmigen Böden.
In Mittel- und Nordeuropa eingeschleppt, ist sie heute fast überall wieder verschwunden.

- **WIRT**

Schmarotzt häuptsächlich auf *Lamium*-Arten. In Südeuropa parasitiert sie hauptsächlich auf verschiedenen Kulturpflanzen (wie zum Beispiel auf *Nicotiana tabacum, Cannabis sativa, Solanum lycopersicum, Zea mays* und *Solanum tuberosum*) und kann dann große Schäden anrichten.

- **GESAMTVERBREITUNG**

Orobanche ramosa besitzt ein großes Verbreitungsgebiet; Mittel- (nördlich bis England, die Niederlande, Norddeutschland, Polen, Estland und Rußland) und Südeuropa; große Teile von Asien, Nord- und Südafrika und Nordamerika.
In Mitteleuropa selten und unbeständig, aber oft in großen Beständen.

- **BEMERKUNGEN**

Orobanche ramosa ist die einzige mitteleuropäische Art, die oft verzweigte (ästige) Blütenstände hat. Ihre Blüten sind viel kleiner als die anderer Arten der Sektion *Trionychon* (*O. arenaria, O. caesia* und *O. purpurea*).

Branched (Hemp) Broomrape

- **SPECIES DESCRIPTION**

The plant is usually slender, approximately 8-45 cm tall. The stem is slender, erect and usually branched, (bright) yellow, yellowish, brown-violet, bluish or pale violet, richly glandular-pubescent, sparsely and laxly scaled. The scale leaves are lanceolate (oval-lanceolate at the base), erect to spreading, dark brown, sparsely glandular-pubescent or glabrous. The inflorescence is usually (very) lax with numerous flowers (spike usually elongated) or cylindrical and relatively dense, with few flowers. The two bracteoles are broadly lanceolate, violet to pale violet or yellowish, about a third of the length of the corolla (as long as or shorter than the calyx-segments) with white glandular hairs. The bract is lanceolate (oval at the base), approximately a third of the size of the corolla (approximately as long as the corolla segments), not deflexed, yellowish-brown or brown and richly pubescent, with white glandular hairs. The calyx is tubular-campanulate, usually 4-dentate and densely glandular-pubescent, about a third of the length of the corolla (calyx-segments shorter than the corolla-tube) and violet to pale violet or (bright) yellow. The flowers are small, erect at first, spreading and curved forward later. The corolla is 10-15 mm long, tubular, slightly constricted just above the ovary, then gradually opening out, conspicuously pubescent, bluish, violet or blue-violet (especially towards the margin), rarely yellow or entirely white with dark violet veins. The dorsal line of the corolla is almost evenly curved from the base onwards, slightly deflexed near the upper lip and sometimes bent upwards at the tip. The upper lip of the corolla is bilobate, pubescent with porrect lobes. The lower lip of the corolla is deflexed and has three oval lobes of almost equal size, with white folds, richly glandular-pubescent. The stamens are inserted 2-4 mm above the base of the corolla-tube. The filaments are almost glabrous or sparsely glandular-pubescent at the base. The anthers are glabrous or with sporadic hairs. The style is sparsely glandular-pubescent. The stigma consists of several spherical lobes and is white or bright bluish, rarely yellowish. $2n = 24$.

- **FLOWERING TIME**

Mid-June to September, from end of March onwards in the Mediterranean region.

- **HABITAT**

Most locations are in sandy fields, especially in root crop cultures, on roadsides, in herbaceous vegetations, in ruderal pastures, on alkaline (frequently calcareous), nutrient-rich sandy or loamy soil.
In the past the plant has been introduced into central and northern Europe, where it now has vanished almost everywhere.

- **HOST**

Mainly parasitic on *Lamium* species. In southern Europe *Orobanche ramosa* is mainly parasitic on various cultivated plants (e.g. *Nicotiana tabacum, Cannabis sativa, Solanum lycopersicum, Zea mays* and *Solanum tuberosum*) and can cause severe damage.

- **DISTRIBUTION**

Orobanche ramosa has a wide range; central (northwards to England, the Netherlands, northern Germany, Poland, Estonia and Russia) and southern Europe; large areas of Asia, northern Africa and South Africa and North America.
In central Europe the plant is rare and fluctuating, but often occurs in large populations.

- **COMMENTS**

Orobanche ramosa is the only central-European species with a branched inflorescence. Its flowers are much smaller than those of other species of the section *Trionychon* (*O. arenaria, O. caesia* and *O. purpurea*).

OROBANCHE RAMOSA

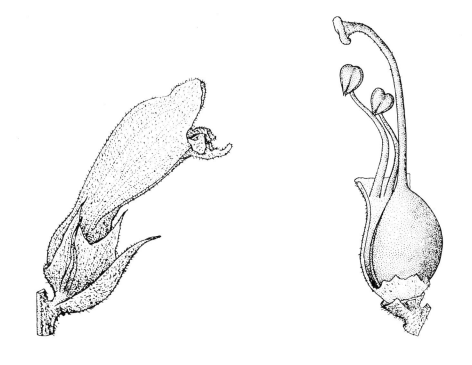

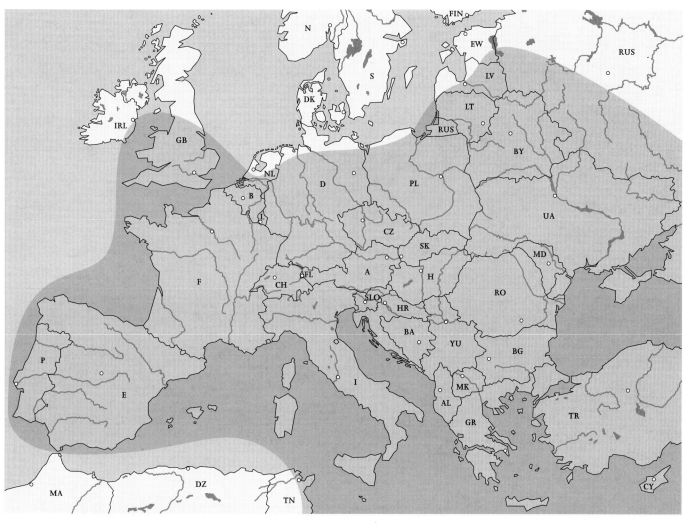

OROBANCHE RAMOSA

Daliyat el Karmil, Har Karmel (IL), 31-3-1992

Villanueva de la Serena, Estremadura (E), 2-4-1994

Gülnar, Icel (TR), 9-5-1988

Setubal, Sierra da Arrabida (P), 22-4-1985

Cala Santanyi, Mallorca (E), 12-4-1988

4.5 SEKTION/SECTION OROBANCHE LINNAEUS 1753

OROBANCHE ALBA — STEPHAN ex WILLDENOW 1800
Orobanche epithymum De Candolle

Weiße (Quendel-) Sommerwurz

- **ARTBESCHREIBUNG**

Die Pflanzen sind meistens schlank und klein, sie erreichen eine Größe von etwa 7 bis 35 cm, selten 60 cm. Der Stengel ist schlank, aufrecht, manchmal schwach gebogen, rötlich überlaufen, selten gelblich gefärbt, reichlich mit Drüsenhaaren besetzt, unten reichlich, oben meist spärlicher beschuppt. Die Schuppen sind eilänglich bis lanzettlich, aufrecht bis abstehend, die unteren kahl, die oberen Schuppen drüsenhaarig. Der Blütenstand ist meist lockerblütig, zylindrisch und mit wenigen Blüten besetzt. Vorblätter sind nicht vorhanden. Das Tragblatt ist etwa so lang wie die Blütenkrone, spärlich mit Drüsenhaaren besetzt, in der Mitte abwärts gekrümmt mit trockener, schwarzbrauner, herabgeschlagener Spitze. Die Kelchhälften sind fast immer ungeteilt, länglich-eiförmig, deutlich nervig, drüsenhaarig, meist ein Drittel bis halb so lang wie die Blütenkrone von gleicher oder dunklerer Farbe als die Blütenkrone. Die Blüten sind ziemlich groß und aufrecht bis abstehend. Die Blütenkrone ist meistens 10 bis 30 mm lang, glockig, über der Ansatzstelle der Staubblätter etwas bauchig erweitert, mit roten oder purpurnen Drüsenhaaren besetzt, außen gelblich bis hellrot, selten weiß mit auffallenden, violetten Adern, innen heller gefärbt. Die Rückenlinie der Blütenkrone ist vom Grund an leicht gekrümmt, in der Mitte leicht oder fast gerade und an der Oberlippe wieder nach vorne gebogen. Die Oberlippe der Blütenkrone ist gekielt, meistens ungeteilt oder ausgerandet, selten zweilappig mit sehr breiten abgerundeten Lappen und reichlich behaart. Der Mittelzipfel der Unterlippe der Blütenkrone ist deutlich länger, fast doppelt so lang wie die Seitenzipfel, am Rande gezähnelt und reichlich mit Drüsenhaaren besetzt. Die Staubblätter sind am oder nahe dem Grund (bis 3 mm hoch) der Kronröhre eingefügt. Die Staubfäden sind an der Basis deutlich behaart, in der Mitte kahl und bis zu den Staubbeuteln mit Drüsenhaaren besetzt. Die Staubbeutel sind an der Naht meistens kahl. Der Griffel ist reichlich dunkel-drüsenhaarig, vor allem in der oberen Hälfte. Die Narbe besteht aus zwei abgerundeten, kugeligen Lappen und ist dunkelrot oder purpurn (selten gelb oder orange) gefärbt. 2n = 38.

- **BLÜTEZEIT**

Mai bis August. Im Gebirge meistens ab Mitte bis Ende Juni.

- **STANDORT**

Die meisten Standorte befinden sich auf warmen Sand-, Trocken- und Halbtrockenrasen (oft an steinigen Stellen), aber auch in Straßenrändern oder buschigen Hängen, auf kalkreichen sandigen Böden.

- **WIRT**

Schmarotzt auf *Lamiaceae* (u.a. auf *Thymus-*, *Salvia-*, *Satureja-* und *Origanum-*Arten), auch auf *Potentilla-*, *Euphorbia-* und *Heracleum-*Arten.

- **GESAMTVERBREITUNG**

Nord-, Mittel- und Südeuropa (vom Atlantischen Ozean bis zum Kaukasus); östlich über Vorderasien bis zum Himalaja, nördlich bis Irland, Schottland, Südbelgien, Mitteldeutschland und Polen. Vom Gebiet abgetrennte Standorte befinden sich auf Gotland und Öland.
Orobanche alba ist ziemlich verbreitet, aber nicht häufig. Sie wächst mit Vorliebe im Bergland und steigt bis in die alpinen Regionen empor.

- **BEMERKUNGEN**

Die Art ist verwechselbar mit *Orobanche teucrii*, aber durch ihre auffallende Farbe der Blumenkrone, ihre deutlich violetten Adern und die Ausbildung der Kelchhälften, die fast immer ungeteilt sind, ist sie von dieser zu unterscheiden.

Thyme (Red) Broomrape

- **SPECIES DESCRIPTION**

The plant is usually small and slender, approximately 7-35 cm tall, rarely 60 cm. The stem is slender, erect, sometimes slightly bent, tinged with red, rarely yellow, richly glandular-pubescent, densely scaled below and usually sparsely scaled above. The scale leaves are oval-elongated to lanceolate, erect to spreading, the lower ones glabrous, the upper ones glandular-pubescent. The inflorescence is usually lax, cylindrical, with few flowers. Bracteoles are absent. The bract is about as long as the corolla, sparsely glandular-pubescent, deflexed in the middle, with a dry brown or black, deflexed tip. The calyx-segments are nearly always entire, oval-elongated, conspicuously veined, glandular-pubescent, usually about a third to half of the length of the corolla and of the same colour as the latter (sometimes darker). The flowers are quite large and erect to spreading. The corolla is usually 10-30 mm long, campanulate, slightly inflated above the insertion of the stamens, with red or purple glandular hairs, yellowish to bright red on the outside, rarely white with conspicuous violet veins; brighter inside. The dorsal line of the corolla is almost evenly curved from the base onwards, almost straight in the middle and bent forward again at the upper lip. The upper lip of the corolla is keeled, usually entire or emarginate, rarely bilobate with very broad, oval lobes and richly pubescent. The central lobe of the lower lip of the corolla is clearly longer, almost twice the length of the left and right lobes, emarginate and richly glandular-pubescent. The stamens are inserted at or near (up to 3 mm) the base of the corolla-tube. The filaments are distinctly pubescent at the base, glabrous in the middle and glandular-pubescent up to the anthers. The anthers are usually glabrous at the line of fusion. The style has many dark glandular hairs, especially in the upper half. The stigma consists of two rounded, spherical lobes and is dark red or purple (rarely yellow or orange). 2n = 38.

- **FLOWERING TIME**

May to August, or from the middle or the end of June in the mountains.

- **HABITAT**

Most locations are in sandy, arid or semi-arid grassland (often on stony ground), but it is also found on roadsides and on wooded slopes on calcareous, sandy soil.

- **HOST**

Parasitic on *Lamiaceae* species (e.g. *Thymus*, *Salvia*, *Satureja*, *Origanum*), as well as on *Potentilla*, *Euphorbia* and *Heracleum* species.

- **DISTRIBUTION**

Northern, central and southern Europe (from the Atlantic to the Caucasus); eastward from Asia Minor to the Himalayas, northwards to Ireland, Scotland, southern Belgium, central Germany and Poland. Isolated locations on Gotland and Öland.
Orobanche alba is quite common, but not frequent. It shows a preference for the mountains and penetrates into alpine regions.

- **COMMENTS**

It is not easy to distinguish *Orobanche alba* from *O. teucrii*, but the conspicuous colour of the corolla of the former, its distinct violet veins and the shape of the corolla segments, which are nearly always entire, allow it to be distinguished from the other species.

OROBANCHE ALBA

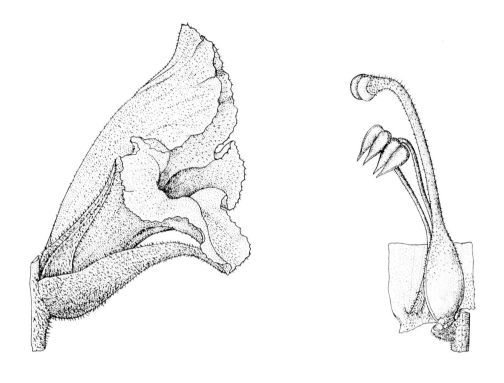

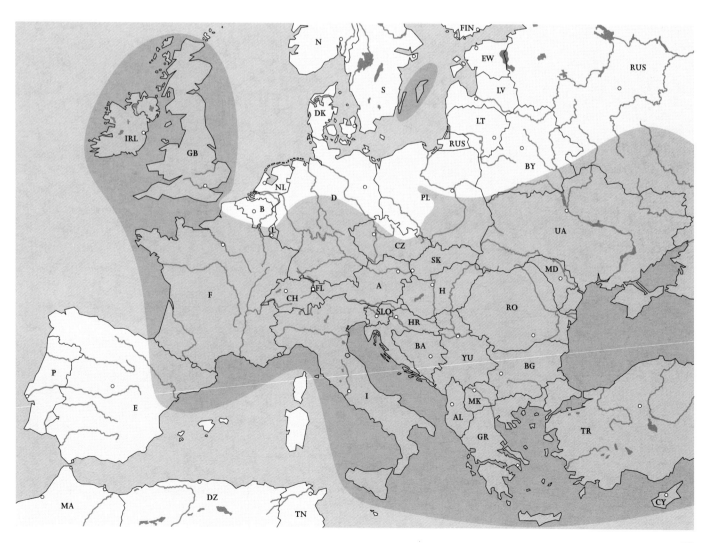

OROBANCHE ALBA

mit Wirtspflanze / with host (*Thymus pannonicus*), Jois, Burgenland (A), 28-5-1994

mit Wirtspflanze / with host (*Thymus pannonicus*), Jois, Burgenland (A), 28-5-1994

mit Wirtspflanze / with host (*Thymus polytrichus*), Tauplitz, Steiermark (A), 25-7-1989

Tauplitz, Steiermark (A), 25-7-1989

Torgny, Luxembourg (B), 26-6-1988

OROBANCHE ALSATICA KIRSCHLEGER 1836

Orobanche cervariae Kirschleger; *O. brachysepala* F.W. Schultz

Elsässer (Haarstrang-) Sommerwurz

- **ARTBESCHREIBUNG**

Die Pflanzen sind meistens kräftig und erreichen eine Größe von etwa (25) 40 bis 70 cm. Der Stengel ist schlank oder kräftig, aufrecht, anfangs gelb, später meist rötlich gefärbt, drüsenhaarig, unten dicht (dachig) beschuppt, im mittleren Teil locker und im oberen Teil spärlich beschuppt. Die Schuppen sind im mittleren und oberen Teil lanzettlich, aufrecht und mit Drüsenhaaren besetzt, die unteren annähernd dreieckig. Der Blütenstand ist meist dicht- und reichlblütig, zylindrisch oder eiförmig mit ziemlich großen Blüten. Vorblätter sind nicht vorhanden. Das Tragblatt ist etwa zwei Drittel so lang wie die Blütenkrone, dunkelbraun gefärbt, drüsenhaarig, manchmal abwärts gebogen, lanzettlich. Die Kelchhälften sind meist ungleich oder fast gleichmäßig zweizähnig, spärlich mit Drüsenhaaren besetzt, genervt und etwa ein Drittel bis halb so lang wie die Blütenkrone und wie diese gefärbt. Die Blüten sind groß, fast waagrecht-abstehend, mit bräunlichlila gefärbten Nerven. Die Blütenkrone ist etwa 20 bis 25 mm lang, über der Ansatzstelle der Staubblätter bauchig erweitert mit weit offenem Schlund mit hellen Drüsenhaaren besetzt, hellgelb bis gelblichbraun, zum Saum hin bräunlichlila gefärbt. Die Rückenlinie der Blütenkrone ist zur Gänze stark und gleichmäßig gebogen (vorn nicht winkelig abfallend) und im Bereich der Oberlippe fast waagrecht. Die Oberlippe der Blütenkrone ist fast ungeteilt oder etwas ausgerandet mit breiten, abstehenden, aufgerichteten Lappen. Die Unterlippe der Blütenkrone besteht aus drei unregelmäßig gezähnelten, herabgebogenen Lappen. Die Staubblätter sind 4 bis 7 mm hoch über dem Grund der Kronröhre eingefügt. Die Staubfäden sind unten stark behaart und oben bis zu den Staubbeuteln spärlich mit Drüsenhaaren besetzt (fast kahl). Die Staubbeutel sind meistens an der Naht behaart. Der Griffel ist reichlich oder spärlich mit Drüsenhaaren besetzt. Die Narbe besteht aus zwei Lappen und ist gelb gefärbt. $2n = 38$.

- **BLÜTEZEIT**

Mitte Juni bis Ende Juli.

- **STANDORT**

Die meisten Standorte befinden sich auf kurzrasigen Trocken- und Halbtrockenrasen in der Nähe lichter Eichen- und Kiefern-Trockenwälder, in Trockengebüschsäumen und in offenen Felsfluren an warmen, sonnigen Standorten auf basenreichen Lehm- und Kalkböden. Zwar wachsen oft viele Exemplare (zerstreut) zusammen, aber es bilden sich selten individuenreiche Gruppen.

- **WIRT**

Vorwiegend auf *Peucedanum cervaria* und *P. alsaticum*.

- **GESAMTVERBREITUNG**

Von Ost-Frankreich über Mitteleuropa bis nach China (nördlich bis Nordostdeutschland, Polen und die baltischen Staaten; südlich bis in den nördlichen Teil der Schweiz und Österreich, ehemaliges Jugoslawien und Rumänien). Auch in Norditalien. Nicht südlich der Alpen, mit Ausnahme vom nördlichen Teil der Balkanhalbinsel. Schwerpunkt der Verbreitung hauptsächlich in Mitteleuropa. Die Art ist sehr selten und tritt zerstreut auf, in vielen Gebieten völlig fehlend.

- **BEMERKUNGEN**

Von der Schwäbischen Alb in Baden-Württemberg und vom Maingebiet bei Karlstadt (Kalbenstein) in Bayern (Deutschland) wurde 1942 durch Suessenguth und Ronniger *Orobanche alsatica* var. *mayeri* beschrieben. Durch K. & F. Bertsch wurde sie 1948 als Art bewertet. Sie unterscheidet sich von *O. alsatica* durch ihre meist reingelben Stengel und Blumenkrone, ihre meistens vorne nicht verwachsenen Kelchhälften, ihre etwas kleineren Blüten (12-15 mm) und ihre Wirtspflanze (*Laserpitium latifolium*).
O. alsatica ist leicht mit *O. elatior* zu verwechseln.

Alsatian Broomrape

- **SPECIES DESCRIPTION**

The plant is usually stout, reaching approximately (25) 40-70 cm. The stem is slender or stout, erect, yellow at first, usually reddish later, glandular-pubescent, densely scaled (imbricate) below, less so in the middle and sparsely scaled above. The scale leaves are lanceolate in the middle and upper parts, erect and glandular-pubescent; lower scale leaves are nearly triangular. The inflorescence is usually dense and many-flowered, cylindrical to oval, with fairly large flowers. Bracteoles are absent. The bract is about two-thirds of the length of the corolla, dark brown, glandular-pubescent, sometimes deflexed, lanceolate. The calyx-segments are usually unevenly or almost regularly bidentate, sparsely glandular-pubescent, veined and about a third or half of the length of the corolla and of the same colour as the latter. The flowers are large, spreading to almost perpendicular, with brownish-violet veins. The corolla is approximately 20-25 mm long, inflated above the insertion of the stamens, with wide open mouth and bright glandular hairs, yellowish to yellowish-brown, brownish-violet near the margin. The dorsal line of the corolla is regularly and strongly curved (not deflexed at the front) and almost horizontal in the range of the upper lip. The upper lip of the corolla is almost entire or slightly emarginate, with broad, porrect, raised lobes. The lower lip of the corolla consists of three irregularly crenate, deflexed lobes. The stamens are inserted 4-7 mm above the base of the corolla-tube. The filaments are distinctly pubescent at the base and sparsely glandular-pubescent (almost glabrous) above, up to the anthers. The anthers are usually pubescent at the line of fusion. The style is richly or sparsely glandular-pubescent. The stigma consists of two yellow lobes. $2n = 38$.

- **FLOWERING TIME**

Mid-June to end of July.

- **HABITAT**

Most locations are in arid and semi-arid low grassland, bordering on open oak forests and dry pine forests, in dry thickets along forests and on open, rocky ground, in warm and sunny places on alkaline loamy and calcareous soil. Although many plants may be found together (though scattered), very large populations are rare.

- **HOST**

Mostly on *Peucedanum cervaria* and *P. alsaticum*.

- **DISTRIBUTION**

From eastern France through central Europe to China (northwards to north-eastern Germany, Poland and the Baltic states; southwards to northern Switzerland and Austria, former Yugoslavia and Rumania). Also in northern Italy. Not south of the Alps, except in the northern Balkans. The core area is central Europe.
Orobanche alsatica is very rare and sporadic; it is absent in many areas.

- **COMMENTS**

Orobanche alsatica var. *mayeri* has been described by Suessenguth and Ronniger in 1942 from the Schwäbische Alb in Baden-Württemberg and from the Main area near Karlstadt (Kalbenstein) in Bavaria (Germany). K. & F. Bertsch regarded it as a species in 1948. It differs from *O. alsatica* in its usually purely yellow stem and corolla, its calyx-segments which are usually not fused at the front, its slightly smaller flowers (12-15 mm) and its host (*Laserpitium latifolium*).
It is difficult to distinguish *Orobanche alsatica* from *O. elatior*.

OROBANCHE ALSATICA

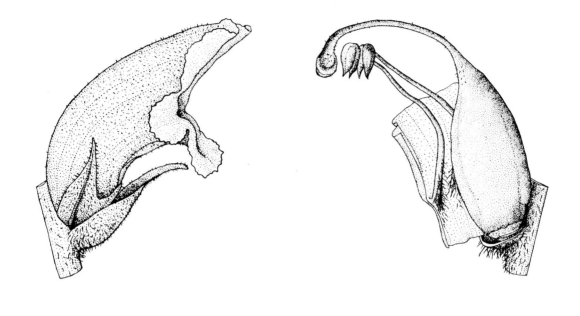

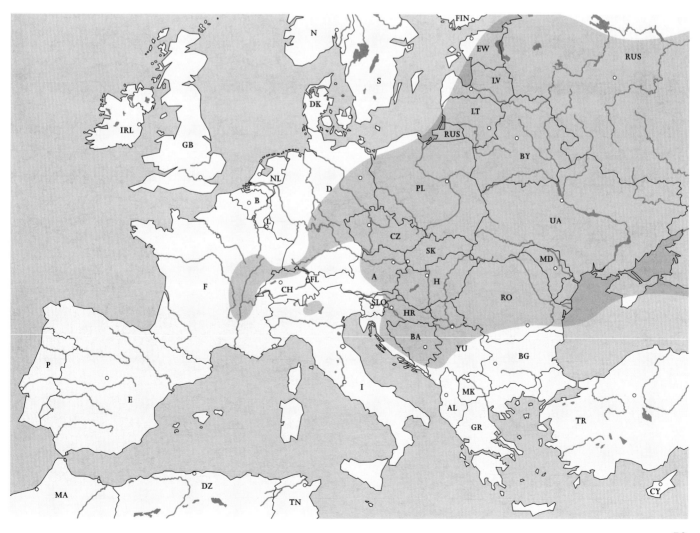

OROBANCHE ALSATICA

Karlstadt, Maintal nördlich Würzburg (D), 23-6-1990

OROBANCHE ALSATICA

mit Wirtspflanze / with host (*Peucedanum cervaria*), Karlstadt, Maintal nördlich Würzburg (D), 23-6-1990

Karlstadt, Maintal nördlich Würzburg (D), 23-6-1990

Karlstadt, Maintal nördlich Würzburg (D), 23-6-1990

Karlstadt, Maintal nördlich Würzburg (D), 23-6-1990

OROBANCHE ALSATICA subsp. MAYERI KIRSCHLEGER (SUESSENGUTH et RONNIGER) C.A.J. KREUTZ

comb. et stat. nov. [Basionym: *Orobanche alsatica* Kirschleger var. *mayeri* Suessenguth et Ronniger; Beitr. Naturk. Forsch. Oberrheingebiet 7: 125-127, 1942]
Orobanche alsatica var. *mayeri* Suessenguth et Ronniger; *O. mayeri* (Suessenguth et Ronniger) K. & F. Bertsch

Mayer's Sommerwurz

• **Artbeschreibung**

Die Pflanzen sind meistens kräftig und erreichen eine Größe von etwa 60 cm. Der Stengel ist schlank oder kräftig, aufrecht, zur Gänze auffallend gelb oder hell- bis dunkelbraun gefärbt, drüsenhaarig, im unteren und mittleren Teil reichlich, im oberen Teil spärlicher beschuppt. Die Schuppen im unteren Teil sind annähernd dreieckig, im mittleren und oberen Teil lanzettlich, aufrecht und mit Drüsenhaaren besetzt. Der Blütenstand ist zylindrisch, reich- und dichtblütig, später vor allem im unteren Teil lockerblütiger und gestreckter mit mittelgroßen Blüten. Vorblätter sind nicht vorhanden. Das Tragblatt ist etwa zwei Drittel so lang wie die Blütenkrone, dunkelbraun gefärbt, lanzettlich, reichlich mit Drüsenhaaren besetzt und in der Mitte abwärts gebogen mit schwarzbrauner Spitze. Die Kelchhälften sind meist ungleich oder fast gleichmäßig zweizähnig, etwa ein Drittel bis halb so lang wie die Blütenkrone, spärlich mit Drüsenhaaren besetzt, heller oder dunkler als diese gefärbt, manchmal mit brauner Spitze. Die vorderen Kelchhälften sind meist nicht verwachsen. Die Blüten sind mittelgroß, aufrecht bis abstehend, oft mit bräunlich gefärbten Nerven. Die Blütenkrone ist etwa 12 bis 15 mm lang, über der Ansatzstelle der Staubblätter bauchig erweitert mit ziemlich weit offenem Schlund mit hellen Drüsenhaaren, hellgelb bis gelblichbraun gefärbt. Die Rückenlinie der Blütenkrone ist zur Gänze gleichmäßig gebogen und im Bereich der Oberlippe fast waagrecht. Die Oberlippe der Blütenkrone ist meist etwas ausgerandet mit breiten, abstehenden, aufgerichteten Lappen. Die Unterlippe der Blütenkrone besteht aus drei unregelmäßig gezähnelten, herabgebogenen Lappen. Die Staubblätter sind 2 bis 4 mm hoch über dem Grund der Kronröhre eingefügt. Die Staubfäden sind am Grunde kahl und oben bis zu den Staubbeuteln spärlich mit Drüsenhaaren besetzt. Die Staubbeutel sind meistens an der Naht behaart. Der Griffel ist im unteren Teil fast ganz kahl und im oberen Teil spärlich mit Drüsenhaaren besetzt. Die Narbe besteht aus zwei Lappen und ist gelb bis orange gefärbt. $2n = 38$.

• **Blütezeit**

Ende Juni bis Ende Juli (meistens Anfang Juli).

• **Standort**

Der *locus classicus* ist eine durch Solitärbuchen aufgelockerte Bergwiese (840-900 m über NN) auf kalkreichem Lehmboden, wobei die nordwestliche Exposition überwiegt. Diese Wiese steht seit 1950 unter Naturschutz und ist einer der wertvollsten Pflanzenstandorte der westlichen Schwäbischen Alb (Deutschland).

• **Wirt**

Nur auf *Laserpitium latifolium*.

• **Gesamtverbreitung**

Nur auf der Schwäbischen Alb in Baden-Württemberg und im Maingebiet bei Karlstadt (Kalbenstein) in Bayern (Deutschland).

• **Bemerkungen**

In der Literatur (unter anderem Suessenguth et Ronniger, 1942) wird fast immer erwähnt, das diese Varietät auffallend gelb gefärbt ist. Am *locus classicus* wachsen jedoch nur einige Pflanzen, die zur Gänze gelb sind, die meisten Pflanzen sind dagegen aber hell- bis dunkelbraun gefärbt. Durch K. & F. Bertsch wurde sie 1948 als Art bewertet.
Orobanche alsatica subsp. *mayeri* unterscheidet sich von *Orobanche alsatica* durch ihren meist reingelben Stengel und Blumenkrone, ihre meistens vorne nicht verwachsenen Kelchhälften, ihre etwas kleineren Blüten (12 bis 15 mm) und einen fast kahlen Griffel.

Mayer's Broomrape

• **Species description**

The plant is usually fairly stout and up to 60 cm tall. The stem is slender or stout, erect and conspicuously yellow or light brown to dark brown over its entire length, glandular throughout, with numerous scale leaves in the lower and middle parts and fewer scale leaves above. The scale leaves are nearly triangular below, lanceolate in the middle and above, erect and glandular. The inflorescence is cylindrical, with numerous flowers in a dense spike, later, mainly in the lower part of the spike, more lax and elongated, with medium-sized flowers. Bracteoles are absent. The bract is approximately two-thirds of the length of the corolla, dark brown, lanceolate, richly glandular, deflexed in the middle, with a brown to black apex. The calyx-segments are usually unequally or almost regularly bifid, about a third to one half of the size of the corolla, sparsely glandular, lighter or darker than the corolla, sometimes with a brown apex. The front halves of the calyx are usually not fused. The flowers are of medium size, erect to erecto-patent, frequently showing brownish veins. The corolla is approximately 12-15 mm long, inflated just above the insertion of the stamens, with a rather wide, funnel-shaped throat with light glandular hairs, bright yellow to yellowish-brown. The dorsal line of the corolla is evenly curved throughout and almost horizontal at the level of the upper lip. The upper lip of the corolla is usually slightly emarginate with broad, spreading, erect lobes. The lower lip of the corolla has three irregularly dentate, deflexed lobes. The stamens are inserted 2-4 mm above the base of the corolla-tube. The filaments are glabrous below and sparsely glandular above, up to the anthers. The anthers are pubescent at the line of fusion. The style is almost glabrous below and sparsely glandular above. The stigma consists of two lobes and is yellow to orange in colour. $2n = 38$.

• **Flowering time**

End of June to end of July (usually early July)

• **Habitat**

The *locus classicus* is a mountain meadow (840-900 m above sea level) on calcareous, loamy soil, with solitary beech trees, with a mainly northwestern exposure. This meadow has been declared a protected nature reserve in 1950 and is one of the most important plant habitats in the western part of the Schwäbische Alb (Germany).

• **Host**

Exclusively on *Laserpitium latifolium*.

• **Distribution**

Found only in the Schwäbisch Alb in Baden-Württemberg and in the Main area near Karlstadt (Kalbenstein) in Bavaria (Germany).

• **Comments**

The literature (Suessenguth et Ronniger, 1942 and others) generally describes the colour of this variety as distinctly yellow. At the *locus classicus*, however, only a few plants are yellow, while most are light to dark brown. K. & F. Bertsch classified the variety as a species in 1948.
Orobanche alsatica subsp. *mayeri* differs from *O. alsatica* by its usually purely yellow stem and corolla, its calyx-segments which are not fused at the front, its smaller flowers (12-15 mm) and an almost glabrous style.

OROBANCHE ALSATICA SUBSP. MAYERI

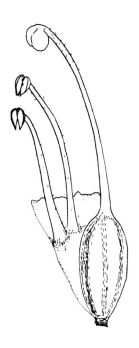

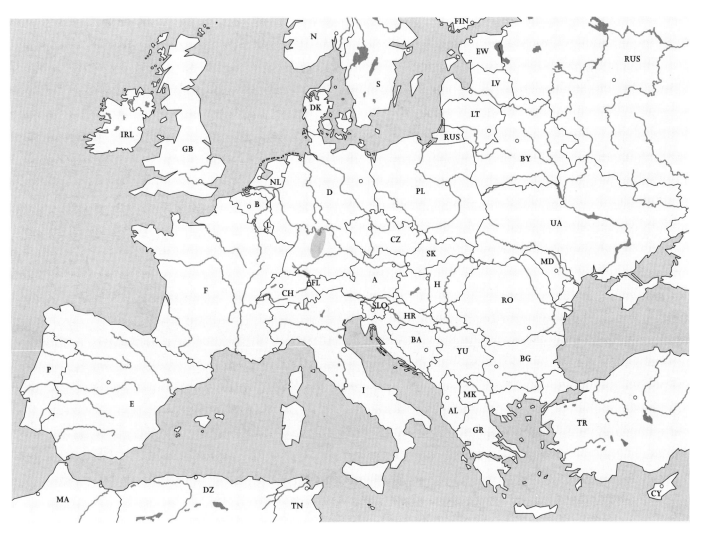

OROBANCHE ALSATICA SUBSP. MAYERI

mit Wirtspflanze / with host (*Laserpitium latifolium*), Hechingen, Schwäbische Alb (D), 6-7-1994

OROBANCHE ALSATICA SUBSP. MAYERI

Hechingen, Schwäbische Alb (D), 6-7-1994

Hechingen, Schwäbische Alb (D), 6-7-1994

Hechingen, Schwäbische Alb (D), 6-7-1994

Hechingen, Schwäbische Alb (D), 6-7-1994

4.8

OROBANCHE AMETHYSTEA — THUILLIER 1797
Orobanche amethystina Reichenbach; *O. eryngii* Duby

Amethyst-Sommerwurz

- **Artbeschreibung**

Pflanzen schlank, meistens kräftig, etwa 15 bis 50 cm hoch. Der Stengel ist kräftig, selten schlank, aufrecht, rotbraun oder violett, selten auch gelblich oder orangegelb gefärbt, drüsenhaarig, unten reichlich, oben meist spärlicher beschuppt. Die unteren Schuppen sind dreieckig bis breit lanzettlich und kahl, die oberen Schuppen sind lanzettlich, drüsenhaarig und abstehend. Der Blütenstand ist meist dicht- und reichblütig, zylindrisch, im unteren Teil später lockerblütiger. Vorblätter sind nicht vorhanden. Das Tragblatt ist lanzettlich, spitz, länger als die Blütenkrone, ab der Mitte abwärts gekrümmt, hell- oder dunkelbraun gefärbt und spärlich mit hellen Drüsenhaaren besetzt. Die Kelchhälften sind ungespalten oder ungleich zweizähnig mit lang zugespitzten Hälften, an der Spitze fast fadenförmig, reichlich mit hellen Drüsenhaaren besetzt, genervt, meist dreiviertel so lang wie die Blütenkrone und dunkler, meist rötlich oder violett wie die Blütenkrone gefärbt. Die Blüten sind mittelgroß, erst aufrecht, später abstehend bis fast waagrecht. Die Blütenkrone ist meistens 12 bis 20 mm lang, röhrig, über der Ansatzstelle der Staubblätter knieförmig oder abwärts gebogen, spärlich drüsenhaarig, gelb, orange- bis goldgelb oder gelblichweiß gefärbt, violett bis lila überlaufen mit dunkelvioletten Nerven. Die Rückenlinie der Blütenkrone ist vom Grund an knieförmig gebogen, in der Mitte fast gerade, im Bereich der Oberlippe abwärts gekrümmt und an der Spitze wenig aufgerichtet. Die Oberlippe der Blütenkrone besteht aus zwei Hälften, mit abstehenden, später zurückgeschlagenen Lappen. Die Unterlippe der Blütenkrone ist herabgeschlagen, violett geädert, mit drei fast gleichgroßen abgerundeten, gezähnelten Lappen. Die Staubblätter sind 3 bis 4,5 mm hoch über dem Grund der Kronröhre eingefügt, am Grund mit Halbmonddrüsen umgeben. Die Staubfäden sind kurz behaart und unter den Staubbeuteln spärlich mit Drüsenhaaren besetzt oder kahl. Die Staubbeutel sind an der Naht spärlich drüsenhaarig. Der Griffel ist spärlich mit Drüsenhaaren besetzt. Die Narbe besteht aus zwei kugeligen Lappen und ist violett, schwarzbraun, rotbraun, selten orange- bis goldgelb gefärbt. $2n = 38$.

- **Blütezeit**

Mai bis Ende Juli.

- **Standort**

Orobanche amethystea ist eine wärmeliebende Pflanze. Die meisten Standorte befinden sich in Trockenrasen, meistens auf südexponierten Hängen, in Xerothermrasen, an lichten Gebüschsäumen auf kalkreichen Lehm- und Lößböden.

- **Wirt**

Schmarotzt vor allem auf *Eryngium campestre*.

- **Gesamtverbreitung**

Mittel- und Südeuropa (vor allem Westeuropa). Von Portugal und Spanien durch Frankreich, Westdeutschland, Italien bis auf die Balkanhalbinsel. Auf Korsika, Sardinien und Sizilien ist sie selten. In Deutschland früher nördlich bis Köln, heute vor allem noch im Rheingebiet. Selten in den ostmediterranen Ländern. Auch in Nordafrika. Angaben für Nordanatolien (die Türkei) sind fraglich (Davis, 1982).
Die Art ist selten und tritt sehr zerstreut auf.

- **Bemerkungen**

Die Pflanzen von Südengland, die durch Philp (1982) zu *Orobanche amethystea* gestellt wurden, gehören zu *O. minor* (Rumsey & Jury, 1991).
Orobanche amethystea ist vor allem an ihrer Blütenfarbe und an ihrem Wirt, in Westeuropa immer *Eryngium campestre*, zu erkennen.

Amethyst Broomrape

- **Species description**

The plant is slender, usually stout, approximately 15-50 cm tall. The stem is stout, rarely slender, erect, red-brown or violet, rarely yellowish or orange-yellow, glandular-pubescent, with numerous scale leaves below and usually with fewer scale leaves above. The lower scale leaves are triangular to broadly lanceolate and glabrous, the upper scale leaves lanceolate, glandular-pubescent and spreading. The inflorescence is cylindrical, with numerous flowers in a dense spike; later on more lax in the lower parts. Bracteoles are absent. The bract is lanceolate, acute, longer than the corolla, deflexed from the middle, bright brown or dark brown and sparsely glandular-pubescent. The calyx-segments are entire or unequally bidentate with long, acuminate halves, almost filiform at the tip, richly glandular-pubescent with light hairs, veined, usually three quarters of the length of the corolla and more darkly coloured than the latter, usually reddish or violet. The flowers are of medium size, erect at first, spreading to almost horizontal later. The corolla is usually 12-20 mm long, tubular, flexed, geniculate or bent downwards, above the insertion of the stamens, sparsely glandular-pubescent, yellow, orange yellow to golden yellow or yellowish-white, tinged with violet, with dark violet veins. The dorsal line of the corolla is curved, geniculate, from the base upwards, almost straight in the middle, bent downwards near the upper lip and slightly raised at the tip. The upper lip of the corolla consists of two halves, with spreading lobes, becoming deflexed later. The lower lip of the corolla has three crenate, oval lobes, which are deflexed and show violet veins. The stamens are inserted 3-4.5 mm above the base of the corolla-tube, with crescent-shaped glands at the base. The filaments are pubescent with short hairs and sparsely glandular-pubescent or glabrous below the anthers. The anthers are sparsely pubescent at the line of fusion. The style is sparsely glandular-pubescent. The stigma consists of two spherical lobes and is violet, brown to black, red-brown, rarely orange yellow to golden yellow. $2n = 38$.

- **Flowering time**

May to end of July.

- **Habitat**

Orobanche amethystea is a thermophile. Most locations are in dry grassland, usually with a southern exposure, in xerothermic grasslands, in open thickets, on calcareous loamy soil and loess.

- **Host**

Parasitic mainly on *Eryngium campestre*.

- **Distribution**

Central and southern Europe (especially in western Europe). From Portugal and Spain, through France, western Germany, Italy to the Balkans. Rare on Corsica, Sardinia and Sicily. In the past northward to Cologne in Germany, now only in the Rhine area. Rare in eastern Mediterranean countries. Also in northern Africa. Findings of *Orobanche amethystea* in northern Anatolia (Turkey) must be considered questionable (Davis, 1982).
Orobanche amethystea is rare and very sporadic.

- **Comments**

The plants in southern England which were classified as *Orobanche amethystea* by Philp (1982) were actually *O. minor* (Rumsey & Jury, 1991).
Orobanche amethystea is recognized mainly by the colour of its flowers and by its host, which in western Europe is always *Eryngium campestre*.

OROBANCHE AMETHYSTEA

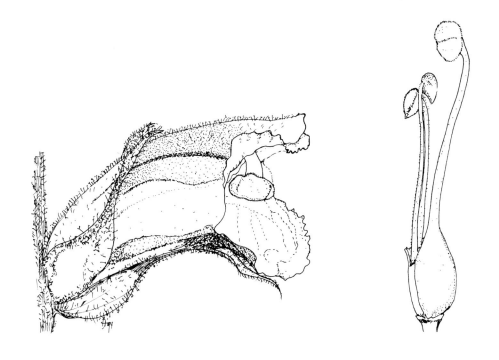

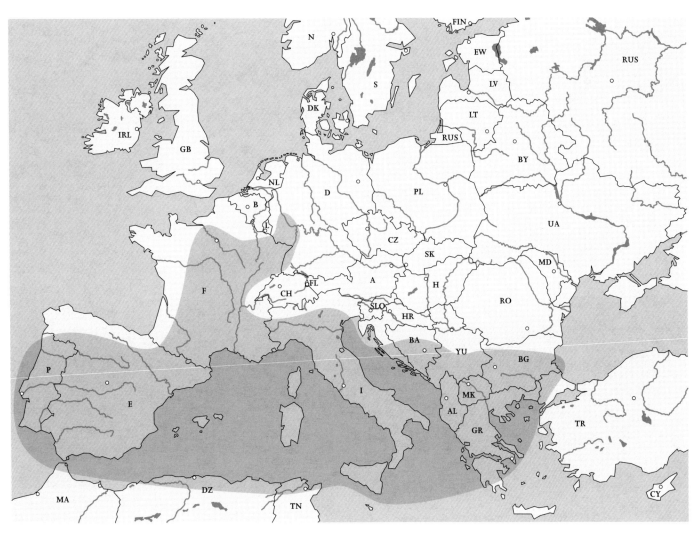

Rouffach, Alsace (F), 22-6-1987

mit Wirtspflanze / with host (*Eryngium campestre*), Orosei, Sardegna (I), 2-5-1990

mit Wirtspflanze / with host (*Eryngium campestre*), Drossohori, Notia Pindos (GR), 17-6-1993

Montboyer, Charente (F), 14-6-1986

Beaufort-sur-Gervanne, Vercors (F), 28-5-1990

4.9

OROBANCHE ARTEMISIAE-CAMPESTRIS — VAUCHER ex GAUDIN 1829
Orobanche artemisiae Vaucher; *O. loricata* Reichenbach

Panzer- (Beifuß-) Sommerwurz

- **ARTBESCHREIBUNG**

Pflanzen schlank bis relativ kräftig, sie erreichen eine Größe von etwa 15 bis 55 cm. Der Stengel ist schlank, machmal kräftig und aufrecht, gelblich, gelblich-braun, bräunlich oder violett-purpurn gefärbt, dicht und wollig behaart, im unteren Teil (Basis) dicht, oben bis zur Hälfte lockerer mit Schuppen besetzt. Die unteren Schuppen sind dreieckig-eiförmig und kahl, die oberen lanzettlich bis länglich und drüsenhaarig, aufrecht. Der Blütenstand ist anfangs meist zylindrisch, dicht- und reichblütig, später langgestreckt und lockerblütig, wobei die Blüten über etwa zwei Drittel am Stengel verteilt sind (Pflanzen mit wenig- und lockerblütigen Ähren sind nicht selten). Vorblätter sind nicht vorhanden. Das schmallanzettliche Tragblatt ist etwas länger als die Blütenkrone, gelblich- bis schwarzbraun gefärbt, fast kahl und ab der Mitte bis zur Spitze abwärts gebogen. Die Kelchhälften sind bis über die Mitte oder fast bis zum Grunde in zwei ungleiche, schmale, an der Spitze fast fadenförmige Zähne gespalten, seltener ungeteilt, die etwa fast gleich lang oder halb so lang wie die Blütenkrone, genervt und meist dunkler (brauner) gefärbt sind. Die Blüten sind mittelgroß, aufrecht bis abstehend. Die Blütenkrone ist etwa 14 bis 22 mm lang, röhrig, über der Ansatzstelle der Staubblätter schwach bauchig erweitert mit außen kurzen hellen Drüsenhaaren, hellgelb bis dunkelgelb gefärbt, mit vor allem zum Saum hin dunkelvioletten Adern. Die Rückenlinie der Blütenkrone ist an der Basis und in der Mitte fast gleichmäßig gebogen (in der Mitte auch fast gerade) und im Bereich der Oberlippe stark nach vorne gekrümmt, der Zipfel der Oberlippe ist wieder aufgerichtet. Die Oberlippe der Blütenkrone ist fast ungeteilt bis zweilappig mit zurückgeschlagenen Zipfeln. Die Unterlippe der Blütenkrone besteht aus drei fast gleichgroßen, rundlichen, gezähnelten, herabgebogenen Lappen. Die Staubblätter sind 3 bis 4 mm hoch über dem Grund der Kronröhre eingefügt. Die Staubfäden sind bis zur Mitte behaart und oben bis zu den Staubbeuteln meist spärlich mit Drüsenhaaren besetzt. Die Staubbeutel sind an der Naht behaart. Der Griffel ist reichlich mit Drüsenhaaren besetzt. Die Narbe besteht aus zwei halbkugeligen Lappen und ist rosa, purpurbräunlich bis rotviolett, im oberen Teil heller als im unteren Teil, gefärbt. 2n = 38.

- **BLÜTEZEIT**

Mitte Juni bis Ende Juli; selten im Mai und manchmal auch noch Anfang August blühend.

- **STANDORT**

Die meisten Standorte befinden sich auf sonnigen, warmen, felsigen, oft südexponierten Hängen, in Xerothermrasen, an lichten Gebüschsäumen, in Trocken- und Halbtrockenrasen auf kalkreichen Lehm- oder Lößboden.

- **WIRT**

Vor allem auf *Artemisia campestris* schmarotzend.

- **GESAMTVERBREITUNG**

Vor allem Süd- und Osteuropa (von Portugal und Südspanien über Südostfrankreich, den südlichen Teil der Schweiz, Norditalien, Niederösterreich, Ostdeutschland, der Tschechischen Republik, den südwestlichen Teil der Slowakei, Ungarn, Rumänien, Nordbulgarien, Ostjugoslawien, Kroatien und Slowenien). Auch in Nordafrika (Marokko).
Die Art ist sehr selten und tritt zerstreut auf.

- **BEMERKUNGEN**

Orobanche artemisiae-campestris ist eine leicht zu erkennende, sehr seltene Art. Sie kann aber mit *O. amethystea* verwechselt werden von der sie sich jedoch durch die dunkelviolette Aderung der Blüten unterscheidet.

Oxtongue Broomrape

- **SPECIES DESCRIPTION**

The plant is slender to relatively stout, approximately 15-55 cm tall. The stem is slender, sometimes stout, erect, yellowish, yellowish-brown, brownish or violet-purple, densely woolly, with numerous scale leaves at the base, fewer above. The lower scale leaves are triangular-oval and glabrous, the higher scale leaves lanceolate to elongated, glandular-pubescent and erect. The inflorescence is usually cylindrical in the beginning, with numerous flowers in a dense spike; elongated and lax later, with the flowers covering two thirds of the stem (specimens with few flowers and a lax spike are fairly common). Bracteoles are absent. The bract is narrow-lanceolate, slightly longer than the corolla, yellowish-brown to dark brown or black, almost glabrous, deflexed from the middle. The calyx-segments are unequally bidentate (divided down to the middle or almost to the base) with narrow halves, almost filiform near the tip, rarely entire, usually nearly equal to or half as long as the corolla, veined and more darkly coloured than the latter, usually brown. The flowers are of medium size, erect to spreading. The corolla is approximately 14-22 mm long, tubular, slightly inflated above the insertion of the stamens, with light glandular hairs on the outside, bright yellow to dark yellow, with dark violet veins, especially near the margin. The dorsal line of the corolla is evenly curved at the base and in the middle (sometimes straight in the middle) and bent conspicuously forward near the upper lip; the tip of the upper lip is raised. The upper lip of the corolla is almost entire to bilobate with recurved tips. The lower lip of the corolla has three crenate, oval lobes, which are deflexed and are almost equal in size. The stamens are inserted 3-4 mm above the base of the corolla-tube. The filaments are pubescent up to the middle and usually sparsely glandular-pubescent above, up to the anthers. The anthers are pubescent at the line of fusion. The style is richly glandular-pubescent. The stigma consists of two hemispherical lobes and is pink, purple-brownish to red-violet, brighter in the upper part. 2n = 38.

- **FLOWERING TIME**

Mid-June to end of July; rarely in May and sometimes in early August.

- **HABITAT**

Most locations are on sunny, warm, rocky slopes with a southern exposure in xerothermic grassland, in open thickets, in arid and semi-arid grassland on calcareous loamy soil or loess.

- **HOST**

Parasitic mainly on *Artemisia campestris*.

- **DISTRIBUTION**

Mainly in southern and eastern Europe (from Portugal and southern Spain, through south-eastern France, southern Switzerland, northern Italy, Niederösterreich (Austria), eastern Germany, the Czech Republic, south-western Slovakia, Hungary, Rumania, northern Bulgaria, eastern Yugoslavia, Croatia and Slovenia). Also in northern Africa (Morocco).
Orobanche artemisiae-campestris is very rare and sporadic.

- **COMMENTS**

Orobanche artemisiae-campestris is easily recognized and very rare. It might be confused with *O. amethystea*, but the dark violet veins of the flowers of the former are characteristic.

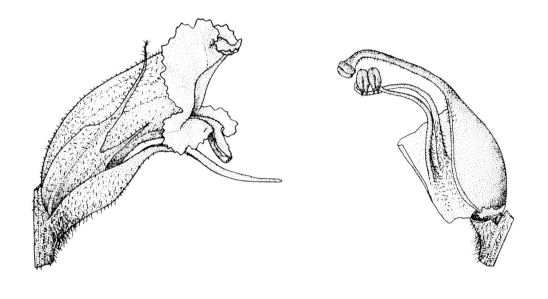
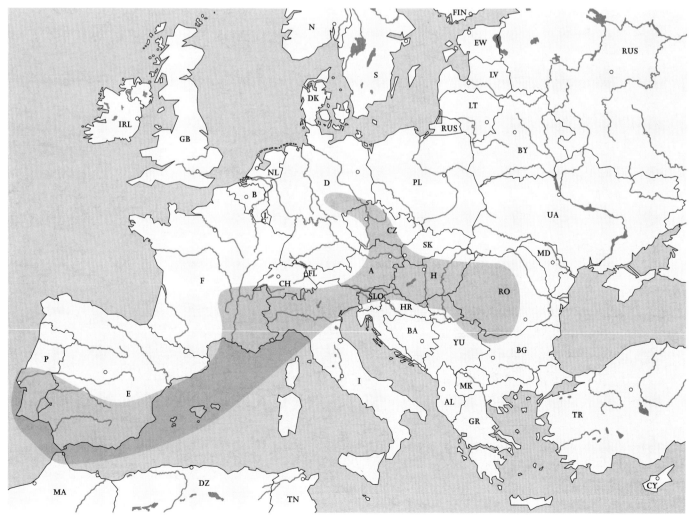

OROBANCHE ARTEMISIAE-CAMPESTRIS

Bad Frankenhausen, Kyffhäusergebirge (D), 16-6-1992

mit Wirtspflanze / with host (*Artemisia campestris*), Bad Frankenhausen, Kyffhäusergebirge (D), 16-6-1992

mit Wirtspflanze / with host (*Artemisia campestris*), Bad Frankenhausen, Kyffhäusergebirge (D), 16-6-1992

Bad Frankenhausen, Kyffhäusergebirge (D), 16-6-1992

Bad Frankenhausen, Kyffhäusergebirge (D), 16-6-1992

4.10

OROBANCHE BARTLINGII — GRISEBACH 1844

Orobanche alsatica var. *libanotidis* (Rupr.) Beck; *O. libanotidis* Rupr.

Bartlings (Heilwurz-) Sommerwurz

• **ARTBESCHREIBUNG**
Die Pflanzen sind meistens schlank und erreichen eine Größe von etwa 15 bis 40 cm. Der Stengel ist schlank, selten kräftig, aufrecht, anfangs gelb, später meist rötlich gefärbt, stark drüsenhaarig, unten dicht, im mittleren und oberen Teil spärlicher beschuppt. Die Schuppen sind im mittleren und oberen Teil eiförmig bis lanzettlich, aufrecht bis abstehend und mit Drüsenhaaren besetzt, die unteren dreieckig, spärlich drüsenhaarig oder kahl. Der Blütenstand ist meist dicht- und reichblütig, zylindrisch oder eiförmig mit kleineren Blüten als *Orobanche alsatica*. Vorblätter sind nicht vorhanden. Das Tragblatt ist etwa zwei Drittel so lang wie die Blütenkrone, dunkelbraun gefärbt, sehr spärlich mit Drüsenhaaren besetzt oder kahl, ab der Mitte abwärts gebogen oder ganz zurückgeschlagen, lanzettlich. Die Kelchhälften sind meist ungleich oder fast gleichmäßig zweizähnig, oft auch vorne verwachsen, meistens etwas heller gefärbt, spärlich mit Drüsenhaaren besetzt, schwach genervt und etwa ein Drittel bis halb so lang wie die Blütenkrone. Die Blüten sind mittelgroß, fast waagrecht abstehend, mit bräunlichlila oder rötlich gefärbten Nerven. Die Blütenkrone ist etwa 12 bis 17 mm lang, über der Ansatzstelle der Staubblätter bauchig erweitert mit weit offenem Schlund und hellen Drüsenhaaren, hellgelb bis gelblichbraun, oder violettrötlich bis rotbraun (rosa) gefärbt. Die Rückenlinie der Blütenkrone ist im unteren Drittel und im Bereich der Oberlippe gleichmäßig gebogen und in der Mitte abgeflacht. Die Oberlippe der Blütenkrone ist fast ungeteilt oder etwas ausgerandet mit breitem, abstehenden, aufgerichteten Lappen. Die Unterlippe der Blütenkrone besteht aus drei unregelmäßig gezähnelten, herabgebogenen Lappen. Die Staubblätter sind 1 bis 3 mm hoch über dem Grund der Kronröhre eingefügt. Die Staubfäden sind unten stark behaart und oben bis zu den Staubbeuteln spärlich mit Drüsenhaaren besetzt (fast kahl). Die Staubbeutel sind meistens an der Naht behaart. Der Griffel ist kahl oder selten am Grunde spärlich mit Drüsenhaaren besetzt. Die Narbe besteht aus zwei Lappen und ist gelb gefärbt. $2n = 38$.

• **BLÜTEZEIT**
Anfang Juni bis Mitte Juli.

• **STANDORT**
Orobanche bartlingii wächst auf Trocken- und Halbtrockenrasen, hauptsächlich in offenen Fluren an steinig-felsigen Kalkhängen an warmen Standorten, aber auch an lichten Gebüschsäumen (Waldränder) auf basenreichen Lehm- und Kalkböden.

• **WIRT**
Nur auf *Seseli libanotis*.

• **GESAMTVERBREITUNG**
Von Mittel- und Osteuropa bis nach China. Schwerpunkt in Osteuropa (Baltische Länder, Rußland bis Sibirien). Südlich bis ehemaliges Jugoslawien und Rumänien. In Deutschland schwerpunktmäßig in den östlichen Bundesländern und in Bayern, viele Standorte befinden sich direkt im ehemaligen DDR/BRD Grenzbereich. Angaben von *Orobanche bartlingii* aus Frankreich beruhen wahrscheinlich auf eine Verwechslung mit *O. alsatica*.
Die Art ist sehr selten und tritt zerstreut auf; in großen Teilen Europas nicht vorhanden.

• **BEMERKUNGEN**
Orobanche bartlingii unterscheidet sich unter anderem von *O. alsatica* durch ihren Wuchs, der zierlicher ist, ihre kleineren Blüten, die Krümmung des Blütenrückens, die Drüsigkeit des Griffels, die Anheftung der Staubblätter und ihre Wirtspflanze (*Seseli libanotis*). Die kontinentale *O. bartlingii* wächst in submontanen, etwas kühleren Lagen mit stärkeren Temperaturschwankungen, während die mitteleuropäische *O. alsatica* ausgesprochene Warmgebiete mit ausgeglichenem Klima bewohnt, wo sie in der unteren, vielfach vom Weinbau eingenommenen, Hügelregion gefunden wird (Nieschalk & Nieschalk, 1974).

Bartling's Broomrape

• **SPECIES DESCRIPTION**
The plant is usually slender, reaching approximately 15-40 cm. The stem is slender, rarely stout, erect, yellow at first, usually reddish later, densely glandular-pubescent, densely scaled below, more sparsely in the middle and above. The scale leaves are oval to lanceolate in the middle and upper parts, erect to spreading and glandular-pubescent; lower scale leaves are triangular and sparsely glandular-pubescent or glabrous. The inflorescence is usually cylindrical or oval, with numerous flowers in a dense spike; the flowers are smaller than those of *Orobanche alsatica*. Bracteoles are absent. The bract is lanceolate, about two thirds of the length of the corolla, dark brown, very sparsely glandular-pubescent or glabrous, deflexed from the middle or completely recurved. The calyx-segments are unequally or almost equally bidentate, often fused at the front, usually lighter in colour, sparsely glandular-pubescent, lightly veined and about a third to half as long as the corolla. The flowers are of medium size, almost horizontal, with brownish-violet or reddish veins. The corolla is approximately 12-17 mm long, slightly inflated above the insertion of the stamens, with a wide funnel-shaped throat and light glandular hairs, bright yellow to yellowish-brown or violet-reddish to red-brown (pink). The dorsal line of the corolla is evenly curved at the base and near the upper lip, straight in the middle. The upper lip of the corolla is almost entire or slightly emarginate, with broad, spreading and erect lobes. The lower lip of the corolla has three irregularly crenate, deflexed lobes. The stamens are inserted 1-3 mm above the base of the corolla-tube. The filaments are richly pubescent below and sparsely glandular-pubescent (almost glabrous) above, up to the anthers. The anthers are usually pubescent at the line of fusion. The style is glabrous or rarely sparsely glandular-pubescent. The stigma consists of two lobes and is yellow. $2n = 38$.

• **FLOWERING TIME**
Early June to mid-July.

• **HABITAT**
Orobanche bartlingii grows in arid and semi-arid grassland, mainly on open ground on stony or rocky, calcareous slopes in warm places, as well as in open thickets (along forests) on alkaline, loamy soil or loess.

• **HOST**
Parasitic only on *Seseli libanotis*.

• **DISTRIBUTION**
From central and eastern Europe to China. Main range is eastern Europe (Baltic states, Russia to Siberia). Southward to former Yugoslavia and Rumania. In Germany mainly in the eastern provinces and Bavaria, many locations are near the former border between DDR and BRD. Findings of *Orobanche bartlingii* in France are probably *O. alsatica*.
Orobanche bartlingii is very rare and sporadic, totally absent in large parts of Europe.

• **COMMENTS**
Orobanche bartlingii can be distinguished from *O. alsatica* by its more graceful build, its smaller flowers, the curve of the dorsal line, its glandular style, the insertion of the stamens and its host, *Seseli libanotis*.
The continental species *O. bartlingii* grows in cool, sub-mountainous regions with more pronounced temperature fluctuations, whereas the central European species *O. alsatica* prefers distinctly warm regions with a more equable climate, and grows at low altitudes in hilly, often wine-growing areas (Nieschalk & Nieschalk, 1974).

OROBANCHE BARTLINGII

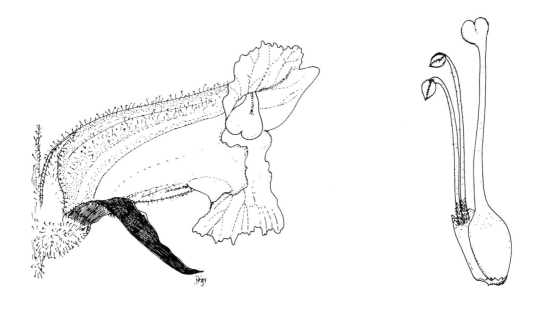

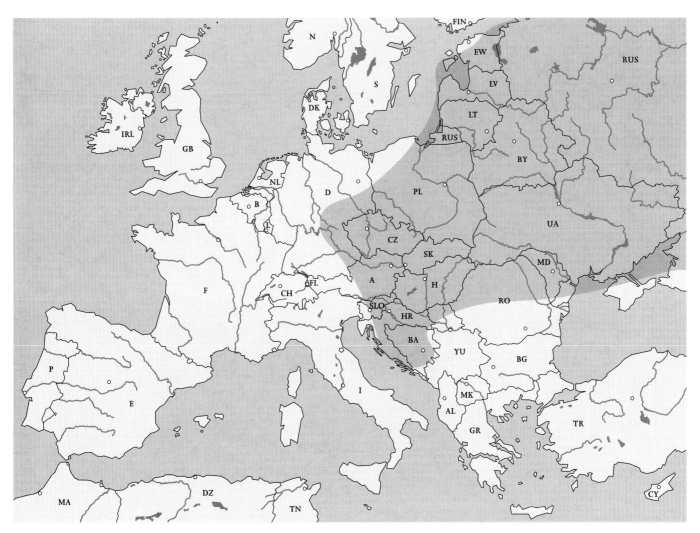

OROBANCHE BARTLINGII

Asbach, Hessisches Bergland (D), 10-7-1994

mit Wirtspflanze / with host (*Seseli libanotis*), Ruprechtsstegen, Nördliche Frankenalb (D), 8-7-1994

mit Wirtspflanze / with host (*Seseli libanotis*), Asbach, Hessisches Bergland (D), 10-7-1994

Ruprechtsstegen, Nördliche Frankenalb (D), 8-7-1994

Allendorf, Hessisches Bergland (D), 21-6-1989

4.11

OROBANCHE CARYOPHYLLACEA ─────────────────────────── SMITH 1798

Orobanche vulgaris Poiret; *O. galii* Duby

Labkraut- (Gemeine) Sommerwurz

• ARTBESCHREIBUNG

Die Pflanzen sind meistens kräftig und können eine Größe van etwa 20 bis 60 cm erreichen. Der Stengel ist schlank, manchmal kräftig und leicht gebogen, blaßgelb bis lila gefärbt, reichlich mit hellen Drüsenhaaren besetzt, unten reichlich, oben meist spärlicher beschuppt. Die Schuppen sind eilänglich bis lanzettlich, aufrecht und mit Drüsenhaaren besetzt. Der Blütenstand ist meist lockerblütig, kurz mit relativ wenigen Blüten. Vorblätter sind nicht vorhanden. Das Tragblatt ist etwa so lang wie die Blütenkrone, rötlichbraun, mit langen hellen Drüsenhaaren besetzt, in der Mitte abwärts gebogen mit trockener, rotbrauner, herabgeschlagener Spitze. Der Kelch besteht aus zwei freien, bis zur Mitte zweispaltigen (selten ungeteilten), eiförmigen, genervten Hälften, reichlich mit hellen Drüsenhaaren besetzt, meist nur halb so lang und von gleicher Farbe wie die Blütenkrone. Die Blüten sind groß und aufrecht-abstehend. Die Blütenkrone ist meistens 15 bis 36 mm lang, vorne deutlich erweitert mit hellen Drüsenhaaren, bräunlich-lila, manchmal bleich oder gelblich, seltener rosa bis rötlich gefärbt. Die Rückenlinie der Blütenkrone ist vom Grund an gekrümmt, in der Mitte leicht oder fast gerade und an der Oberlippe etwas nach vorne gebogen. Die Oberlippe der Blütenkrone ist sehr breit, gekielt, schwach ausgerandet mit erst aufgerichteten, dann vorgestreckten Lappen. Die Unterlippe der Blütenkrone ist etwas herabgeschlagen mit drei fast gleichgroßen, längsfaltigen, gezähnelten Lappen. Die Staubblätter sind am oder nahe dem Grund (bis 3 mm hoch) der Kronröhre eingefügt. Die Staubfäden sind unten verbreitert, bis zur Mitte dicht behaart und ab der Mitte bis zu den Staubbeuteln reichlich oder spärlich mit Drüsenhaaren besetzt. Die Staubbeutel sind kahl oder an der Naht warzig behaart. Der Griffel ist an der Basis spärlich behaart und in der oberen Hälfte reichlich mit Drüsenhaaren besetzt. Die Narbe besteht aus zwei abgerundeten Lappen und ist braunviolett bis dunkelpurpurn gefärbt. Die Art duftet stark nach Nelken. 2n=38.

• BLÜTEZEIT

Mai bis August. Eine frühblühende Art, blüht in den niedrigen Lagen meistens schon ab Mitte Mai. Im Gebirge meistens ab Mitte bis Ende Juni.

• STANDORT

Die meisten Standorte befinden sich auf Halbtrockenrasen und in Wiesen, aber auch im Wald und an Straßenrändern (seltener im Gebüsch) auf lockeren, kalkhaltigen Lehm- oder Lößböden in warmen Lagen.

• WIRT

Schmarotzt vor allem auf *Galium*- (*Galium mollugo* und *G. verum*) und auf *Asperula*-Arten.

• GESAMTVERBREITUNG

Von Mittel- und Osteuropa (nordwärts bis Südostengland, den Niederlanden, Deutschland, Polen und Estland) bis nach Vorderasien, den Kaukasus und Iran. In den Nordalpen und im nördlichen Alpenvorland fast nur in den Föhntälern. In den südlichen Ländern, welche dem mediterranen Florengebiete angehören, ist sie seltener und wächst hier meistens im Mittelgebirge. In Südnorwegen adventiv. Auch in Nordafrika (Meusel *et al.*, 1978).
Die Art ist weit verbreitet und in manchen Gebieten nicht selten.

• BEMERKUNGEN

Orobanche caryophyllacea ist an ihrem meistens leicht gebogenen Stengel, ihrem auffälligen Nelkengeruch (der weniger ausgeprägt nur noch bei *O. gracilis* vorhanden ist), ihrer hellbraunvioletten Blütenfarbe, ihrem lockeren Blütenstand und ihren großen Blüten leicht zu erkennen.

Bedstraw (Clove-scented) Broomrape

• SPECIES DESCRIPTION

The plant is usually stout, reaching approximately 20-60 cm. The stem is slender, sometimes stout and slightly curved, pale yellow to violet, richly glandular-pubescent, densely scaled below, usually sparsely scaled above. The scale leaves are oval-elongated to lanceolate, erect and glandular-pubescent. The inflorescence is usually short, with relatively few flowers in a lax spike. Bracteoles are absent. The bract is about as long as the corolla, reddish-brown, pubescent with long, light, glandular hairs, deflexed in the middle, with a shrivelled, red-brown, recurved tip. The calyx consists of two oval, veined halves, divided down to the middle (rarely entire), richly glandular-pubescent and usually about half as long as the corolla; it has the same colour as the latter. The flowers are large and erecto-patent. The corolla is usually 15-36 mm long, distinctly widening towards the mouth, with light glandular hairs, brownish-violet, sometimes pale or yellowish, rarely pink or reddish. The dorsal line of the corolla is evenly curved from the base, less curved or straight in the middle and bent forward near the upper lip. The upper lip of the corolla is very broad, keeled, slightly emarginate, its lobes erect at first and porrect later. The lower lip of the corolla is slightly deflexed and has three crenate, plicate lobes of equal size. The stamens are inserted near or at the base of the corolla-tube (at a height of up to 3 mm). The filaments are swollen at the base, richly pubescent up to the middle and sparsely or richly glandular-pubescent above. The anthers are glabrous or have warty hairs at the line of fusion. The style is sparsely pubescent at the base and richly glandular-pubescent from the middle up. The stigma consists of two oval lobes and is brown-violet to dark purple. *Orobanche caryophyllacea* emits a strong carnation scent. 2n = 38.

• FLOWERING TIME

May to August. The plant flowers early in the year, at low altitudes usually from mid-May onwards. Usually from the middle to the end of June in the mountains.

• HABITAT

Most locations are in semi-arid grasslands and meadows, also in forests and on road-sides (rarely under shrubs) on loose, calcareous, loamy soil and loess, in warm places.

• HOST

Parasitic mainly on *Galium* (*Galium mollugo* and *G. verum*) and on *Asperula* species.

• DISTRIBUTION

From central and eastern Europe (northward to south-eastern England, the Netherlands, Germany, Poland, Estonia) to Asia Minor, the Caucasus and Iran. In the northern alps and in the northern alpine foothills almost exclusively in Föhn-valleys. In southern countries which are part of the Mediterranean floral region *Orobanche caryophyllacea* grows mainly in sub-alpine areas. In southern Norway adventive. Also in northern Africa (Meusel *et al.*, 1978).
The plant has a wide range and is not rare in many areas.

• COMMENTS

Orobanche caryophyllacea is easily recognised by its slightly curved stem, its conspicuous carnation scent (the only other *Orobanche* species with this scent, albeit much weaker, is *O. gracilis*), its bright brown-violet flowers, its lax inflorescence and its large flowers.

OROBANCHE CARYOPHYLLACEA

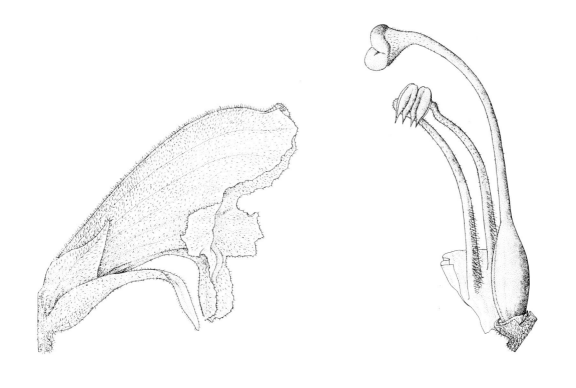

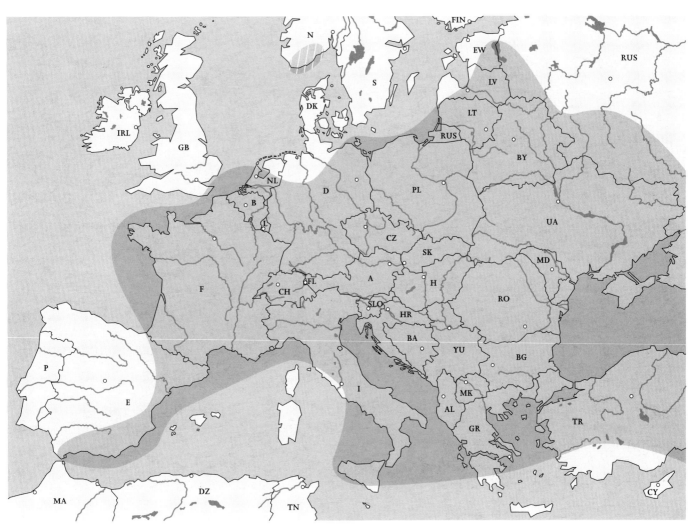

mit Wirtspflanze / with host (*Galium mollugo*), Jois, Burgenland (A), 29-5-1986

Nonnenbach, Eifel (D), 2-7-1989

Villardelle, Aude (F), 22-5-1990

Rouffach, Alsace (F), 22-6-1987

Le Flon, Lac de Taney (CH), 4-7-1992

4.12

OROBANCHE CERNUA

LOEFLING 1758

Nickende Sommerwurz

- **Artbeschreibung**

Pflanzen ziemlich kräftig, etwa 10 bis 25 cm hoch. Der Stengel ist kräftig bis sehr kräftig, aufrecht, blau, bläulichviolett, gelb bis rötlich gefärbt, spärlich drüsenhaarig oder fast kahl, bis zum Blütenstand locker und spärlich beschuppt. Die Schuppen sind breiteiförmig bis eilanzettlich (kurz) und spärlich mit Drüsenhaaren besetzt, aufrecht bis abstehend (die Spitze oft aufrecht). Der Blütenstand ist meist dicht-, reichblütig und zylindrisch, später lockerblütiger und gestreckt. Vorblätter sind nicht vorhanden. Das Tragblatt ist etwa halb oder ein Drittel so lang wie die Blütenkrone, eiförmig-lanzettlich, goldgelb, bräunlichgelb (an der Basis heller gefärbt), drüsenhaarig und nicht herabgeschlagen. Die Kelchhälften sind in zwei tief ungleiche Zähne gespalten, selten aus ungeteilten Hälften bestehend (auf der unteren Seite miteinander verwachsen), eiförmig oder eilanzettlich, ein Drittel bis etwa halb so lang wie die Blütenkrone, gelb, goldgelb oder violett bis bläulich gefärbt, schwach genervt, mit kurzen und hellen Drüsenhaaren besetzt. Die Blüten sind klein bis mittelgroß, anfangs aufrecht-abstehend, später vorwärts bis abwärts (geknickt) gebogen. Die Blütenkrone ist meistens 12 bis 15 mm lang, über der Ansatzstelle der Staubblätter bauchig erweitert, in der Mitte verengt und bogig nach vorne gekrümmt (kniefömig) und im Bereich der Oberlippe wieder schwach bauchig erweitert, sehr spärlich drüsenhaarig oder fast kahl, weiß und gegen den Saum zu (blau)violett gefärbt. Die Rückenlinie der Blütenkrone ist im unteren Drittel stark nach vorne gebogen (geknickt) und im mittleren und oberen Teil gleichmäßig, schwach unregelmäßig gebogen oder fast gerade, oft mit einem aufgerichteten Spitzchen. Die Oberlippe der Blütenkrone besteht aus zwei gezähnelten, aufrecht abstehenden Lappen mit fast kahlen Rändern. Die Unterlippe der Blütenkrone ist schwach herabgeschlagen oder vorgestreckt mit drei fast gleichgroßen, gerundeten oder spitzlichen Lappen. Die Staubblätter sind 4 bis 6 mm hoch über dem Grund der Kronröhre eingefügt. Die Staubfäden sind am Grunde kahl oder behaart und oben bis zu den Staubbeuteln meist spärlich drüsenhaarig. Die Staubbeutel sind an der Naht spärlich behaart oder kahl. Der Griffel ist spärlich mit Drüsenhaaren besetzt oder kahl. Die Narbe besteht aus zwei verlängerten, kugeligen Lappen, die weiß bis gelblichweiß gefärbt sind. $2n = 38$.

- **Blütezeit**

Juni bis Ende Juli, im mediterraner Bereich schon ab Ende März.

- **Standort**

In Xerotherm- und Trockenrasen, steinigen büschigen Abhängen, in ruderal beeinflußten Halbtrockenrasen; im Küstenbereich oft in Sanddünen auf trockenen, lockeren, sandigen Böden. Die Art wächst vor allem an sonnigen und warmen Standorten Südeuropas.

- **Wirt**

Schmarotzt hauptsächlich auf *Artemisia*-Arten.

- **Gesamtverbreitung**

Vor allem das westliche mediterrane Gebiet von Europa (nördlich bis Südfrankreich und Süditalien). Weiter östlich über Bulgarien, Vorder- und Zentralasien bis nach China. Nicht auf Sardinien und den Balearen. Auch in Ostindien, Nordafrika und Australien. Der nördlichste Standort befindet sich in Südtirol in Italien.
In fast allen Teilen Europas sehr selten; in den Mittelmeerländern tritt sie manchmal häufig auf.

- **Bemerkungen**

Orobanche cernua ist eine leicht zu bestimmende Art. Sie ist vor allem an ihrer geknickten Rückenlinie und ihrer weißen Blütenfarbe mit bläulichviolettem Saum zu erkennen. Ihre Blüten sind kleiner als die Blüten von *O. cumana*.

Nodding Broomrape

- **Species description**

The plant is quite robust, approximately 10-25 cm tall. The stem is robust to very robust, erect, blue, bluish-violet, yellow to reddish, sparsely glandular-pubescent or glabrous, laxly and sparsely scaled up to the inflorescence. The scale leaves are broadly oval to oval-lanceolate (short) and sparsely glandular-pubescent, erect to spreading (the tip is often erect). The inflorescence is usually cylindrical with numerous flowers in a dense spike, later more lax and elongated. Bracteoles are absent. The bract is about a third to half as long as the corolla, oval-lanceolate, golden yellow, brownish-yellow (lighter at the base), glandular-pubescent and not deflexed. The calyx-segments are deeply bifid, with two unequal teeth, rarely consisting of two entire segments (connate at the base), oval or oval-lanceolate, about one third to half as long as the corolla, yellow, golden yellow or violet to bluish, slightly veined, glandular-pubescent with short, light glandular hairs. The flowers are small to medium-sized, erecto-patent at first, bent forward and downward (nodding) later. The corolla is usually 12-15 mm long, inflated above the insertion of the stamens, constricted and arched forward (geniculate) in the middle, slightly inflated again near the upper lip, very sparsely glandular-pubescent or almost glabrous, white, tinged with (bluish) violet near the margin. The dorsal line of the corolla is strongly curved forward (geniculate) in the lower part, evenly curved (with some irregularities) or almost straight in the middle and upper parts, often with a raised tip. The upper lip of the corolla consists of two denticulate, erecto-patent lobes with an almost glabrous margin. The lower lip of the corolla is slightly deflexed or porrect with three rounded or acute lobes of almost equal size. The stamens are inserted 4-6 mm above the base of the corolla-tube. The filaments are glabrous or pubescent at the base and sparsely glandular-pubescent above, up to the anthers. The anthers are sparsely pubescent or glabrous at the line of fusion. The style is sparsely pubescent or glabrous. The stigma consists of two elongated, spherical lobes which are white to yellowish-white. $2n = 38$.

- **Flowering time**

June to end of July, as early as end of March in the Mediterranean region.

- **Habitat**

In xerothermic and dry grassland, on stony and shrubby slopes, in ruderal semi-arid grassland; near the coast often on sand dunes on dry, loose, sandy soil. *Orobanche cernua* grows mainly in sunny and warm places in southern Europe.

- **Host**

Parasitic mainly on *Artemisia* species

- **Distribution**

Mainly in the western Mediterranean region of Europe (northward to the South of France and southern Italy), further eastward through Bulgaria, Asia Minor and central Asia to China. Not on Sardinia and the Balearic Islands. Also in eastern India, northern Africa and Australia. The northernmost location is Alto Adige (Italy).
Orobanche cernua is very rare in most parts of Europe; sometimes frequent in Mediterranean countries.

- **Comments**

Orobanche cernua is easily recognised by its inflected dorsal line and its colours, the flower being white with a bluish-violet margin. Its flowers are smaller than those of *O. cumana*.

OROBANCHE CERNUA

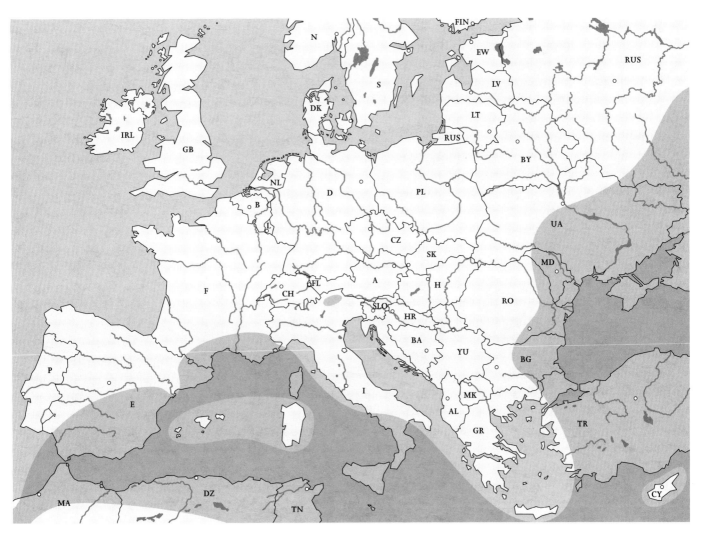

OROBANCHE CERNUA

Cabo de Gata, Sierra de Gata (E), 18-4-1994

mit Wirtspflanze / with host (*Artemisia barrelieri*), Cabo de Gata, Sierra de Gata (E), 18-4-1994

Cabo de Gata, Sierra de Gata (E), 18-4-1994

Netanya, Ha-Sharon (IL), 5-4-1992

En Gedi, Ha-Negev (IL), 14-3-1992

4.13

OROBANCHE COERULESCENS — STEPHAN ex WILLDENOW 1800

Bläuliche Sommerwurz

- **ARTBESCHREIBUNG**

Pflanzen kräftig und niedrig, etwa 7 bis 30 cm hoch. Der Stengel ist kräftig und sehr dick (vor allem im unteren Teil), aufrecht, im unteren Teil gelblichbraun, im oberen Teil hellgelb oder gelblich gefärbt, unten fast kahl, oben reichlich weißwollig behaart und bis zum Blütenstand ziemlich reichlich beschuppt. Die unteren Schuppen sind eilanzettlich (dreieckig) bis lanzettlich und fast kahl, die oberen schmallanzettlich und in unteren Teil weißwollig behaart, aufrecht bis abstehend. Der Blütenstand ist meist dichtblütig, zylindrisch, später im unteren Teil lockerblütiger. Vorblätter sind nicht vorhanden. Das Tragblatt ist etwas kürzer als die Blütenkrone, weißwollig behaart, breit-eiförmig, schwarzbraun (an der Basis gelblich), im oberen Teil mit herabgeschlagener Spitze. Die Kelchhälften sind aus zwei getrennten, zweispaltigen, seltener dreispaltigen Hälften gebildet, eilanzettlich, wollig behaart, etwa ein Drittel bis halb so lang wie die Blütenkrone, etwas heller oder gleich wie die Blütenkrone gefärbt. Die Blüten sind klein, anfangs aufrecht-abstehend, später vorwärts gebogen bis waagrecht abstehend. Die Blütenkrone ist meistens 15 bis 18, selten bis 20 mm lang, über der Ansatzstelle der Staubblätter verengt (stark eingeschnürt), und gegen den Saum wieder erweitert, an der Außenseite reichlich mit hellen (weißen) Drüsenhaaren besetzt, hellblau oder blauviolett (am Grunde weißlich) gefärbt. Die Rückenlinie der Blütenkrone ist von der Basis stark vorwärts gebogen und im mittleren und oberen Teil fast gerade. Die Oberlippe der Blütenkrone ist vorgestreckt und besteht aus zwei kurzen, breiten, kahlen, runden Lappen, die am Rande schwach gekerbt sind. Die Unterlippe der Blütenkrone ist ein wenig herabgeschlagen oder vorgestreckt, mit drei fast gleichgroßen (der mittlere ist etwas länger), gezähnelten, abgerundeten Lappen, die alle mit großen weißen Falten versehen sind. Die Staubblätter sind 6 bis 7 mm hoch über dem Grund der Kronröhre eingefügt. Die Staubfäden sind am Grunde schwach behaart, in der Mitte und oben bis zu den Staubbeuteln kahl. Die Staubbeutel sind an der Naht meistens fast kahl. Der Griffel ist schwach behaart oder kahl; der Fruchtknoten hat Längsfurchen. Die Narbe ist weiß oder gelblichweiß gefärbt. 2n = 38.

- **BLÜTEZEIT**

Juni und Juli. Die einzelnen Pflanzen haben eine relativ lange Blütezeit.

- **STANDORT**

Auf Trocken-, (Halbtrocken-), Steppen-, und Xerothermrasen, an lichten Gebüschsäumen (Waldrändern), in Felsfluren an warmen, sonnigen Standorten auf basenreichen, sandigen Böden.

- **WIRT**

Schmarotzt vorwiegend auf *Artemisia campestris*, seltener auf *Achillea millefolium*.

- **GESAMTVERBREITUNG**

Von Südostbayern über Ostdeutschland, nördlich bis Lettland (Riga), über das östliche Europa durch Zentralasien bis Japan. Hauptverbreitung in den Steppengebieten von Eurasien. Nicht in den Mittelmeerländern. Am westlichen Arealrand ist die Art vielfach erloschen.
In Europa ist *Orobanche coerulescens* sehr selten und nur noch auf wenige Standorte beschränkt; in den Steppen Zentralasiens ist sie eine häufige Erscheinung.

- **BEMERKUNGEN**

In Mitteleuropa kommt nur die Form '*occidentalis*' G. Beck 1890 vor, die kleinere Blüten besitzt.
Orobanche coerulescens unterscheidet sich unter anderem von den anderen Arten durch ihre hellblaue Blütenfarbe und ihren kurzen, aber sehr dicken und kräftigen Stengel.

Bluish Broomrape

- **SPECIES DESCRIPTION**

The plant is robust and low, approximately 7-30 cm tall. The stem is robust and very thick (especially below), erect, yellowish-brown below, light yellow or yellowish above, almost glabrous below, richly woolly with white hairs above and quite densely scaled, up to the inflorescence. The lower scale leaves are oval-lanceolate (triangular) to lanceolate and almost glabrous, the higher scale leaves are narrowly lanceolate, woolly with white hairs in the lower part and erect to spreading. The inflorescence is usually cylindrical and dense, later more lax in the lower part. Bracteoles are absent. The bract is slightly shorter than the corolla, woolly with white hairs, broadly oval, brown to black (yellowish at the base), with a deflexed tip. The calyx-segments consist of two separate, bifid or trifid halves, oval-lanceolate, woolly pubescent, about one third or half as long as the corolla, slightly lighter than or with the same colour as the latter. The flowers are small, erecto-patent at first, bent forward or horizontal later. The corolla is usually 15-18 mm long, rarely longer than 20 mm, (distinctly) constricted above the insertion of the stamens, widening again near the margin, richly glandular-pubescent with light (white) glandular hairs on the outside, light blue or blue-violet (whitish at the base). The dorsal line of the corolla is strongly curved forward from the base and almost straight in the middle and upper parts. The upper lip of the corolla is porrect and consists of two short, broad, glabrous, oval lobes with a slightly crenate margin. The lower lip of the corolla is slightly deflexed or porrect with three, crenate, oval lobes of almost equal size (the middle lobe being slightly longer), with conspicuous white folds. The stamens are inserted 6-7 mm above the base of the corolla-tube. The filaments are sparsely pubescent at the base and glabrous in the middle and above, up to the anthers. The anthers are usually almost glabrous at the line of fusion. The style is sparsely pubescent or glabrous; the ovary has longitudinal grooves. The stigma is white to yellowish-white. 2n = 38.

- **FLOWERING TIME**

June and July. Individual plants have a relatively long flowering period.

- **HABITAT**

In arid or semi-arid grasslands, steppes, xerothermic grasslands, in open thickets (along forests) and on rocky ground, in warm, sunny places on alkaline, sandy soil.

- **HOST**

Parasitic mainly on *Artemisia campestris*, more rarely on *Achillea millefolium*.

- **DISTRIBUTION**

From south-eastern Bavaria, through eastern Germany, northward to Lithuania (Riga), through eastern Europe and central Asia to Japan. Its main range is the steppes of Eurasia. Not in the Mediterranean countries. Mostly extinct at the western border of the range.
Orobanche coerulescens is very rare in Europe and limited to a few places; it is frequent in the steppes of central Asia.

- **COMMENTS**

Central Europe has only the form '*occidentalis*' G. Beck 1890, which has smaller flowers.
Orobanche coerulescens is easily recognized by its bright blue flowers and its short, but very thick and robust stem.

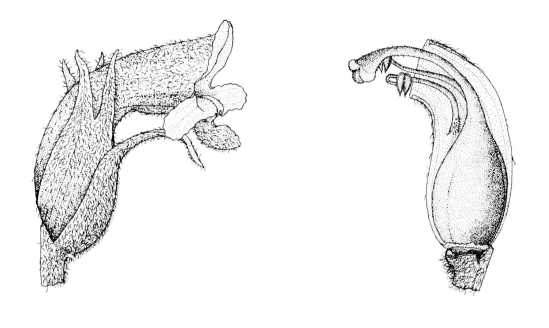
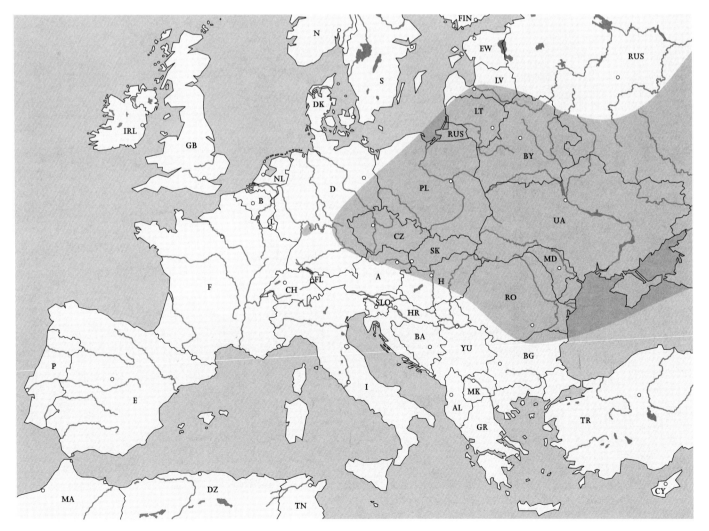

OROBANCHE COERULESCENS

mit Wirtspflanze / with host (*Artemisia campestris*), Rothenbruck, Nördliche Frankenalb (D), 7-7-1994

OROBANCHE COERULESCENS

mit Wirtspflanze / with host (*Artemisia campestris*), Rothenbruck, Nördliche Frankenalb (D), 7-7-1994

Rothenbruck, Nördliche Frankenalb (D), 7-7-1994

Pfaffenhofen, Nördliche Frankenalb (D), 2-7-1993

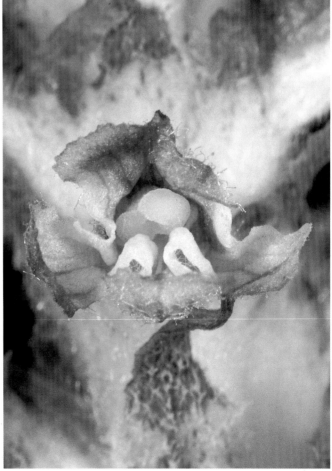

Litomerice, Severocesky (CZ), 25-6-1993

4.14

OROBANCHE CRENATA ——————————————————————— FORSKAL 1775

Orobanche speciosa De Candolle; *O. pruinosa* Lapeyrouse; *O. segetum* Spruner; *O. klugei* Schmitz et Regel; *O. picta* Wilms; *O. pelargonii* Caldesi

Gezähnelte Sommerwurz

- ARTBESCHREIBUNG

Pflanzen meist kräftig bis sehr kräftig, etwa 15 bis 50, auch bis fast 100 cm hoch. Der Stengel ist kräftig, selten schlank, aufrecht, rotbraun, gelblichweiß, goldgelb, gelbbraun, rötlich oder rötlichbraun gefärbt, reichlich mit hellen und kurzen Drüsenhaaren besetzt oder fast kahl, bis zum Blütenstand spärlich und locker beschuppt. Die Schuppen sind eilänglich bis lanzettlich und fast kahl, aufrecht bis abstehend. Der Blütenstand ist sehr dicht-, reichblütig und zylindrisch oder sehr locker-, reichblütig und gestreckt (im oberen Teil zylindrisch und dichtblütig bleibend), wobei die untersten Blüten oft weit voneinander entfernt und weit unten am Stengel stehen. Vorblätter sind nicht vorhanden. Das Tragblatt ist etwa so lang wie die Blütenkrone, lanzettlich, schwarzbraun gefärbt, im oberen Teil mit herabgeschlagener Spitze, drüsenhaarig. Die Kelchhälften sind zweispaltig, ungleich zweizähnig oder voneinander getrennt, schmallanzettlich bis fadenförmig, genervt, fast kahl, dreiviertel bis etwa gleich lang wie die Blütenkrone, gelblichweiß mit violettem Saum oder zur Gänze violett gefärbt. Die Blüten sind mittelgroß bis groß, anfangs aufrecht, später mehr abstehend. Die Blütenkrone ist meistens 25 bis 30 (bei kleineren Exemplaren auch 10 bis 20) mm lang, röhrig-glockig erweitert, über der Ansatzstelle der Staubblätter bauchig erweitert, spärlich mit hellen Drüsenhaaren besetzt oder fast kahl, weiß (bleich) oder gelblichweiß gefärbt, besonders im Bereich der Oberlippe violett oder rosa überlaufen; die ganze Blüte ist schwach oder stark violett geädert. Die Rückenlinie der Blütenkrone ist an der Basis stark gebogen, in der Mitte fast gerade oder schwach gebogen und kurz vor der Oberlippe ein wenig aufgerichtet. Die Oberlippe der Blütenkrone ist abgerundet, ausgerandet, mit sehr breiten vorgestreckten, abstehenden oder zurückgebogenen Lappen; ihre Zipfel sind kahl. Die Unterlippe der Blütenkrone besteht aus drei tief gezähnelten, gerundeten und gefalteten Lappen, wobei der Mittellappen oft größer als die beiden Seitenlappen ist. Die Staubblätter sind 2 bis 5 mm hoch über dem Grund der Kronröhre eingefügt. Die Staubfäden sind am Grunde reichlich behaart und oben bis zu den Staubbeuteln spärlich behaart, seltener kahl. Die Staubbeutel sind an der Naht oft behaart. Der Griffel ist mit Drüsenhaaren besetzt. Die Narbe besteht aus zwei verlängerten, kugeligen Lappen, die weiß, orange, gelblichweiß, rosa oder hellviolett gefärbt sind. Die Blüten riechen angenehm nach Nelken. 2n = 38.

- BLÜTEZEIT

Mai bis Ende Juli, im mediterranen Bereich schon ab Mitte April.

- STANDORT

Die meisten Standorte befinden sich an Straßenrändern, in Unkrautgesellschaften, in ruderal beeinflußten Wiesen und an Ackerrändern auf basen- (kalkhaltigen), nährstoffreichen, lehmigen Böden.
Orobanche crenata parasitiert in den Mittelmeerländern vor allem in landwirtschaftlichen Kulturen und kann dort sehr große Schäden verursachen.

- WIRT

Schmarotzt hauptsächlich auf *Fabaceae*-, seltener auf *Pelargonium*-Arten, manchmal auch auf Arten anderer Familien. Bei den Kulturpflanzen bevorzugt sie vor allem *Vicia faba*, *V. ervilia*, *Daucus carota*, *Cicer arietinum*, *Pisum sativum* und *Lens culinaris*.

- GESAMTVERBREITUNG

Südeuropa; von Portugal, Spanien, Südfrankreich, Italien, Griechenland und der Türkei bis in die Kaukasusländer. Auch in Nordafrika und auf den Kanarischen Inseln.
In Mittel- und Nordeuropa (unter anderem Estland, der Tschechischen Republik, die Niederlande, Nordostfrankreich, Südengland, die Schweiz, Norditalien, Mittel- und Süddeutschland und Niederösterreich) ist *Orobanche crenata* vorübergehend eingeschleppt worden.

- BEMERKUNGEN

Orobanche crenata ist leicht zu erkennen. Die Pflanzen sind sehr kräftig, die Blüten meistens weiß mit violetten Adern. Sie schmarotzt vor allem in landwirtschaftlichen Kulturen.

Carnation-scented (Bean) Broomrape

- SPECIES DESCRIPTION

The plant is usually stout to very stout, approximately 15-50 or even up to 100 cm tall. The stem is stout, rarely slender, erect, red-brown, yellowish-white, golden yellow, reddish or reddish-brown, richly glandular-pubescent with light, short hairs or almost glabrous, laxly and sparsely scaled up to the inflorescence. The scale leaves are oval-elongated to lanceolate and almost glabrous, erect to spreading. The inflorescence usually has numerous flowers in a dense, cylindrical spike or a lax, elongated spike with numerous flowers (remaining cylindrical and dense above) with the lowest flowers often far apart and very low on the stem. Bracteoles are absent. The bract is as long as the corolla, lanceolate, dark-brown, glandular-pubescent, deflexed at the tip. The calyx-segments are bifid or unequally bidentate or totally separated, narrowly lanceolate to filiform, veined, almost glabrous, about two thirds of or as long as the corolla, yellowish-white with a violet margin or violet all over. The flowers are medium-sized to large, erect at first, more spreading later. The corolla is usually 25-30 mm (10-20 mm in small plants) long, tubular-campanulate, inflated above the insertion of the stamens, sparsely glandular-pubescent with light glandular hairs or almost glabrous, white (pale) or yellowish-white; tinged with violet or pink, especially near the upper lip; the entire flower has light or dark violet veins. The dorsal line is clearly curved forward at the base, almost straight or slightly curved in the middle, often erect near the tip. The upper lip of the corolla is rounded, emarginate, with very broad, porrect, spreading or deflexed lobes; the tips of the lobes are glabrous. The lower lip of the corolla consists of three deeply crenate, rounded and plicate lobes, the middle lobe usually being larger than the side lobes. The stamens are inserted 2-5 mm above the base of the corolla-tube. The filaments are richly pubescent at the base and sparsely pubescent up to the anthers, rarely glabrous. The anthers are often pubescent at the line of fusion. The style is glandular-pubescent. The stigma consists of two elongated, spherical lobes, which are white, orange, yellowish-white, pink or light violet. The flowers emit a pleasant scent of carnations. 2n = 38.

- FLOWERING TIME

May to end of July, as early as mid-April in the Mediterranean region.

- HABITAT

Most locations are on roadsides, in herbaceous vegetations, in ruderal pastures and on borders of arable fields on alkaline (calcareous), nutrient-rich loamy soil. *Orobanche crenata* is parasitic mainly on agricultural plants in the Mediterranean countries and causes extensive damage to crops.

- HOST

Parasitic mainly on *Fabaceae* species, more rarely on *Pelargonium* species, sometimes on species of other genera. Among cultivated plants *Orobanche crenata* prefers *Vicia faba*, *V. ervilia*, *Daucus carota*, *Cicer arietinum*, *Pisum sativum* and *Lens culinaris*.

- DISTRIBUTION

Southern Europe; from Portugal, Spain, southern France, Italy, Greece and Turkey to the Caucasus. Also in northern Africa and on the Canary Islands.
The plant has been introduced temporarily into central and northern Europe (e.g. in Estonia, the Czech Republic, the Netherlands, Northeastern part of France, southern England, Switzerland, northern Italy, central and southern Germany and Niederösterreich (Austria)).

- COMMENTS

Orobanche crenata is easily recognized. The plant is very stout, flowers are usually white with violet veins. It is found mainly in arable fields.

OROBANCHE CRENATA

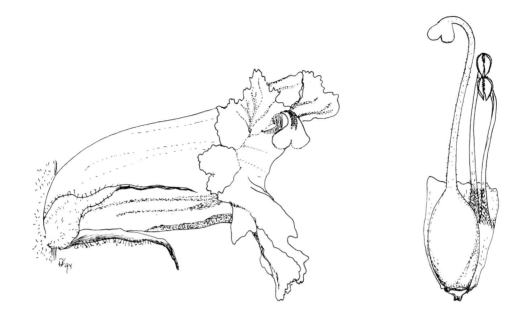

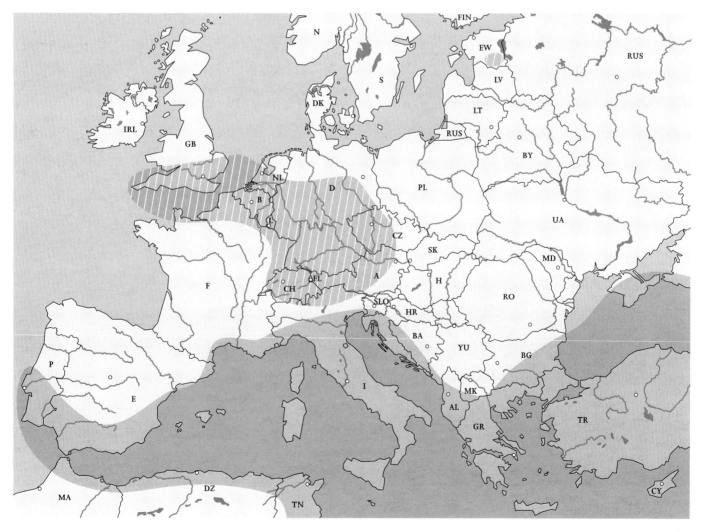

OROBANCHE CRENATA

Villanueva de la Serena, Estremadura (E), 2-4-1994

OROBANCHE CRENATA

mit Wirtspflanze / with host (*Daucus carota*), Doshen, Jordan (IL), 4-4-1992

Villanueva de la Serena, Estremadura (E), 2-4-1994

Kusadasi, Izmir (TR), 5-5-1988

Cala Santanyi, Mallorca (E), 12-4-1988

4.15

OROBANCHE CUMANA ─────────────────────────── WALLROTH 1825

Sonnenblumen-Sommerwurz

- **ARTBESCHREIBUNG**

Pflanzen schlank, etwa 10 bis 50 cm hoch. Der Stengel ist schlank, aufrecht (selten leicht gebogen), bläulichviolett, gelblich bis rötlich gefärbt, spärlich drüsenhaarig oder fast kahl, locker und spärlich beschuppt. Die Schuppen sind breiteiförmig bis eilanzettlich (kurz) und spärlich mit Drüsenhaaren besetzt, aufrecht. Der Blütenstand ist meist locker- und reichblütig, später im unteren Teil sehr lockerblütig, wobei die untersten Blüten oft weit voneinander entfernt und weit unten, öfter ganz unten am Stengel stehen. Vorblätter sind nicht vorhanden. Das Tragblatt ist etwa ein Drittel, halb oder fast so lang wie die Blütenkrone, eiförmig-lanzettlich, bräunlichgelb oder braunviolett (an der Basis heller gefärbt), drüsenhaarig, an der Spitze herabgeschlagen. Die Kelchhälften sind in zwei tief ungleiche Zähne gespalten, selten aus ungeteilten Hälften bestehend (auf der unteren Seite miteinander verwachsen), eiförmig oder eilanzettlich, ein Drittel bis etwa halb so lang wie die Blütenkrone, violett, braun oder bläulich gefärbt, mit kurzen, hellen Drüsenhaaren und genervt. Die Blüten sind klein bis mittelgroß, anfangs aufrecht-abstehend, später vorwärts bis abwärts (geknickt) gebogen. Die Blütenkrone ist meistens 15 bis 18 mm lang, über der Ansatzstelle der Staubblätter nicht erweitert, in der Mitte verengt und bogig nach vorne gekrümmt (knieförmig) und im Bereich der Oberlippe schwach bauchig erweitert, sehr spärlich drüsenhaarig oder fast kahl, weiß und gegen den Saum zu (blau)violett gefärbt, vor allem der Rücken ist bläulich oder violett überlaufen. Die Rückenlinie der Blütenkrone ist im unteren Drittel nach vorne gebogen (geknickt), im mittleren Teil gleichmäßig oder schwach unregelmäßig gebogen oder fast gerade und im Bereich der Oberlippe wieder schwach abwärts gebogen, oft mit einem aufgerichteten Spitzchen. Die Oberlippe der Blütenkrone besteht aus zwei gezähnelten, aufrecht abstehenden Lappen mit kahlen Rändern. Die Unterlippe der Blütenkrone ist herabgeschlagen oder vorgestreckt mit drei fast gleichgroßen, gerundeten oder spitzlichen Lappen. Die Staubblätter sind 4 bis 6 mm hoch über dem Grund der Kronröhre eingefügt. Die Staubfäden sind am Grunde kahl oder behaart und oben bis zu den Staubbeuteln meist spärlich drüsenhaarig. Die Staubbeutel sind an der Naht spärlich behaart oder kahl. Der Griffel ist spärlich mit Drüsenhaaren besetzt oder kahl. Die Narbe besteht aus zwei verlängerten, kugeligen, weiß oder weißlich (und violett überlaufenen) gefärbten Lappen. $2n = 38$.

- **BLÜTEZEIT**

Juni bis Ende Juli, im mediterranen Bereich schon ab Ende März.

- **STANDORT**

Fast ausschließlich in Sonnenblumenkulturen auf kalkhaltigen Lehm- oder Lößböden.

- **WIRT**

Schmarotzt hauptsächlich auf *Helianthus annuus*. Sie parasitiert vor allem in den Mittelmeerländern und in Südosteuropa zu Tausenden in den Sonnenblumenkulturen (auf einer Wirtspflanze manchmal bis zu 86 Exemplare) und fügt der Landwirtschaft große Schäden zu.

- **GESAMTVERBREITUNG**

Fast das ganze mediterrane Gebiet von Europa (nördlich bis Südfrankreich, Süditalien, die Slowakei und die Ukraine) bis nach China, besonders in Vorder- und Zentralasien. Nicht auf Sardinien, Zypern und den Balearen. Auch in Nordafrika und Australien.
In Mitteleuropa teilweise eingeschleppt und sehr zerstreut.

- **BEMERKUNGEN**

Orobanche cumana wächst fast ausschließlich in Sonnenblumenkulturen. Sie besitzt einen schlankeren Habitus als *O. cernua*, außerdem ist ihr Blütenstand sehr locker und lang und ihre Blüten sind größer als 15 mm.

Sunflower Broomrape

- **SPECIES DESCRIPTION**

The plant is slender, approximately 10-50 cm tall. The stem is slender, erect (rarely slightly curved), bluish-violet, yellow or reddish, sparsely glandular-pubescent or almost glabrous, laxly and sparsely scaled. The scale leaves are broadly oval to oval-lanceolate (short), erect and sparsely glandular pubescent. The inflorescence usually has numerous flowers in a lax spike, later very lax below, with the flowers far apart and very low, often right down to the base of the stem. Bracteoles are absent. The bract is about one third to one half or even as long as the corolla, oval-lanceolate, brownish-yellow or brown-violet (lighter at the base), glandular-pubescent and deflexed at the tip. The calyx-segments are deeply bifid, with two unequal teeth, rarely consisting of two entire halves (connate in the lower half), oval or oval-lanceolate, about one third to half as long as the corolla, violet, brown or bluish, veined and glandular-pubescent with short, light glandular hairs. The flowers are small to medium-sized, erect and spreading at first, horizontal or nodding later. The corolla is usually 15-18 mm long, not inflated above the insertion of the stamens, constricted in the middle and bent forwards (geniculate), then slightly inflated near the upper lip, very sparsely glandular-pubescent or almost glabrous, white and blue (-violet) near the margin, its back tinged with blue or violet. The dorsal line is curved forward (inflected) in the lower third of the corolla, evenly or unevenly curved or almost straight in the middle; curved downward again near the upper lip, often with a raised tip. The upper lip of the corolla consists of two denticulate, erecto-patent lobes with a glabrous margin. The lower lip of the corolla is deflexed or porrect and has three rounded or acute lobes of almost equal sizes. The stamens are inserted 4-6 mm above the base of the corolla-tube. The filaments are glabrous or pubescent at the base and sparsely glandular-pubescent up to the anthers. The anthers are glabrous or sparsely pubescent at the line of fusion. The style is sparsely glandular-pubescent or glabrous. The stigma consists of two, elongated, spherical lobes, which are white or whitish (and tinged with violet). $2n = 38$.

- **FLOWERING TIME**

June to end of July, as early as end of March in the Mediterranean region.

- **HABITAT**

Almost exclusively in sunflower cultures on calcareous, loamy soil or loess.

- **HOST**

Parasitic mainly on *Helianthus annuus*. Thousands of specimens can be found together in sunflower cultures (up to 86 parasites on a single host plant) in the Mediterranean countries, where they cause extensive damage to crops.

- **DISTRIBUTION**

Almost the entire Mediterranean area of Europe (northward to southern France, southern Italy, the Slovak Republic and the Ukraine) to China, especially in Asia Minor and central Asia. Not on Sardinia, Cyprus or on the Balearic Islands. Also in northern Africa and Australia.
Sporadically introduced into central Europe.

- **COMMENTS**

Orobanche cumana grows almost exclusively in sunflower cultures. Its appearance is more slender than that of *O. cernua* and its inflorescence is very lax and elongated, with flowers exceeding 15 mm.

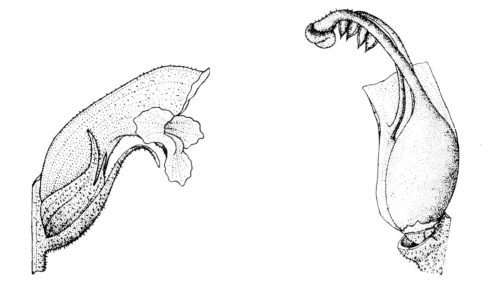

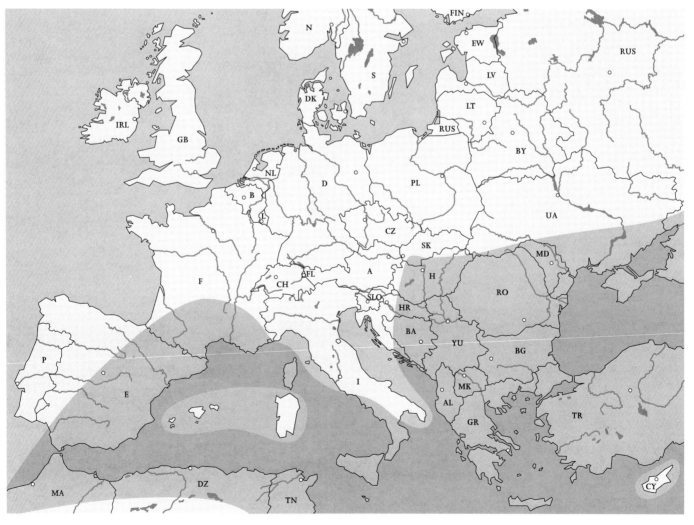

OROBANCHE CUMANA

Fuente de Piedra, Llanos de Antequera (E), 26-4-1994

mit Wirtspflanze / with host (*Helianthus annuus*), Bet She'an, Haré Gilboa (IL), 22-5-1984

Bet She'an, Haré Gilboa (IL), 22-5-1984

Fuente de Piedra, Llanos de Antequera (E), 26-4-1994

Fuente de Piedra, Llanos de Antequera (E), 26-4-1994

OROBANCHE ELATIOR — SUTTON 1797

Orobanche major L. 1753; *O. centaureae scabiosae* F. Schultz; *O. fragrans* Koch; *O. stigmatodes* Wimmer; *O. kochii* F. Schultz

Große Sommerwurz

- **ARTBESCHREIBUNG**

Die Pflanzen sind groß und (sehr) kräftig und erreichen eine Größe von etwa 20 bis 70 cm. Der Stengel ist kräftig, seltener schlank, und aufrecht, braun bis dunkel(rot-)braun, gelblich oder rosa gefärbt, meistens reichlich drüsenhaarig und gleichmäßig und reichlich bis zum Blütenstand beschuppt. Die Schuppen sind lanzettlich, im unteren Teil dreieckig-eiförmig, aufrecht und mit Drüsenhaaren besetzt. Der Blütenstand ist meist dichtblütig, zylindrisch und reichblütig. Vorblätter sind nicht vorhanden. Das Tragblatt ist etwa so lang wie die Blütenkrone, sehr spärlich drüsenhaarig oder fast kahl, ab der Mitte abwärts gebogen, rötlich- bis schwarzbraun gefärbt, lanzettlich mit trockener, schwarzbrauner, herabgeschlagener Spitze. Die Kelchhälften bestehen aus zwei sich vorne berührenden oder verwachsenen, ungleich zweizähnigen Hälften, etwa ein Drittel bis halb so lang wie die Blütenkrone, oft etwas heller als die Blütenkrone gefärbt und reichlich mit Drüsenhaaren besetzt. Die Blüten sind relativ groß, zuerst aufrecht-abstehend, später waagrecht-abstehend, mit dunkler gefärbten Nerven. Die Blütenkrone ist etwa 15 bis 28 mm lang, über der Ansatzstelle der Staubblätter allmählich erweitert mit hellen Drüsenhaaren, erst gelbbraun mit rötlichem Anflug, später gelblich, braungelblich oder rotbraungelb, selten zur Gänze mit kräftiger gelber Färbung. Die Rückenlinie der Blütenkrone ist vom Grund an gleichmäßig gebogen, in der Mitte manchmal etwas stärker nach vorne gekrümmt. Die Oberlippe der Blütenkrone ist meist ungeteilt oder etwas ausgerandet mit zwei rundlichen, aufgerichteten Lappen. Die Unterlippe der Blütenkrone besteht aus drei fast gleichgroßen gezähnelten, herabgebogenen Lappen. Die Staubblätter sind 4 bis 6 mm hoch über dem Grund der Kronröhre eingefügt, am Grund mit goldgelbem Nektarfleck. Die Staubfäden sind unten an der Vorderseite dicht behaart und oben bis zu den Staubbeuteln spärlicher mit Drüsenhaaren besetzt. Die Staubbeutel sind oft an der Naht behaart. Der Griffel ist vor allem im oberen Teil reichlich mit Drüsenhaaren besetzt. Die Narbe besteht aus zwei Lappen und ist gelb gefärbt. $2n = 38$.

- **BLÜTEZEIT**

Juni bis August. Die Hauptblütezeit liegt in der zweiten Juni-Hälfte.

- **STANDORT**

Die meisten Standorte befinden sich auf kurzrasigen Trocken- und Halbtrockenrasen, in Trockengebüschsäumen und in trockenen Fettwiesen an sonnigen Stellen auf basenreichen Lehm- und Lößböden.

- **WIRT**

Schmarotzt vor allem auf *Centaurea scabiosa*, selten auch auf anderen *Centaurea*-Arten. Auch auf andere *Asteraceae*- und auf *Thalictrum*-Arten.

- **GESAMTVERBREITUNG**

Zerstreut vom Atlantischen Ozean durch Europa bis Zentralasien (Himalaya) und Nordindien. Nordwärts bis England, Dänemark, Südschweden und die baltischen Staaten. Südlich bis Nordostspanien, Mittelitalien und Südgriechenland (Peloponnes). Im vielen Ländern des Mittelmeergebietes fehlt diese Art oder ist sie sehr selten. In den Niederlanden erstmals 1973 gefunden (Kreutz, 1988, 1989).
Die Art ist selten, aber in manchen Gebieten tritt sie häufig auf.

- **BEMERKUNGEN**

Orobanche elatior ist meist nur schwer von *O. lutea* zu unterscheiden. Man achte dazu auf die verschiedenen Merkmale der beiden Arten (vor allem auf die spätere Blütezeit). *O. elatior* ist meistens kräftiger als *O. lutea*.

Tall (Knapweed) Broomrape

- **SPECIES DESCRIPTION**

The plant is tall and (very) stout, reaching approximately 20-70 cm. The stem is stout, rarely slender, erect, brown to dark (reddish) brown, yellowish or pink, usually richly glandular-pubescent, evenly and densely scaled up to the inflorescence. The scale leaves are lanceolate, triangular lanceolate below, erect and glandular-pubescent. The inflorescence usually has numerous flowers in a dense, cylindrical spike. Bracteoles are absent. The bract is about as long as the corolla, very sparsely glandular-pubescent or almost glabrous, deflexed from the middle, reddish-brown to dark brown or black, lanceolate with a shrivelled, dark brown, recurved tip. The calyx-segments consist of two unequally bidentate halves, touching or fused at the base, about a third to half the length of the corolla and often of a somewhat lighter colour than the latter, richly glandular-pubescent. The flowers are quite large, erect and spreading at first, horizontal later, with dark veins. The corolla is about 15-28 mm long, gradually widening above the insertion of the stamens, with light glandular hairs, yellowish-brown with a tinge of red at first, yellowish, brown-yellowish or red-brown-yellow later, rarely entirely bright yellow. The dorsal line of the corolla is evenly curved from the base onwards, sometimes more clearly bent forward in the middle. The upper lip of the corolla is usually entire or slightly emarginate, with two rounded, erect lobes. The lower lip of the corolla consists of three crenate, deflexed lobes of almost equal size. The stamens are inserted 4-6 mm above the base of the corolla-tube; they have a golden yellow nectar-spot at the base. The filaments are densely pubescent at the front and more sparsely glabrous-pubescent up to the anthers. The anthers are often glabrous at the line of fusion. The style is richly glandular-pubescent, especially above. The stigma consists of two lobes and is yellow in colour. $2n = 38$.

- **FLOWERING TIME**

June to August. The main flowering time is the second half of June.

- **HABITAT**

Most locations are in arid and semi-arid grassland (short turf), in dry thickets and in dry, fertile meadows in sunny places on alkaline loamy soil and loess.

- **HOST**

Parasitic mainly on *Centaurea scabiosa*, rarely on other *Centaurea* species. Also on other *Asteraceae* species and on *Thalictrum* species.

- **DISTRIBUTION**

Scattered from the Atlantic, through Europe to central Asia (Himalayas) and northern India. Northward to England, Denmark, southern Sweden, and the Baltic states. Southward to north-eastern Spain, central Italy and southern Greece (Peloponnesus). Very rare or totally absent in many countries of the Mediterranean. First found in the Netherlands in 1973 (Kreutz, 1988, 1989).
This species is rare, but frequent in some areas.

- **COMMENTS**

It is difficult to distinguish *Orobanche elatior* from *O. lutea*. Attention should be paid to the different characteristics of both species (in particular to the later flowering time). *O. elatior* is usually more stoutly built than *O. lutea*.

OROBANCHE ELATIOR

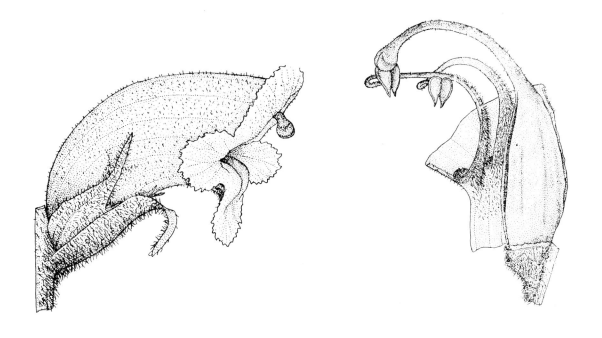

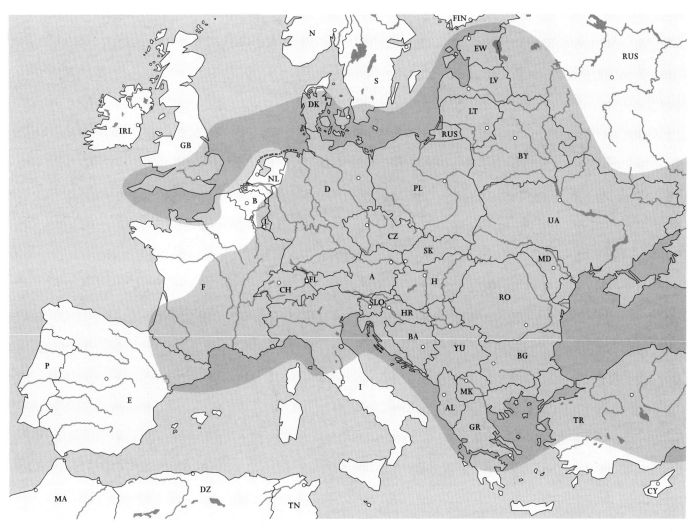

OROBANCHE ELATIOR

Nonnenbach, Eifel (D), 8-7-1988

mit Wirtspflanze / with host (*Centaurea scabiosa*), Nonnenbach, Eifel (D), 8-7-1988

Voerendaal, Zuid-Limburg (NL), 21-6-1988

Vogtsburg, Kaiserstuhl (D), 21-6-1987

Vogtsburg, Kaiserstuhl (D), 21-6-1987

4.17

OROBANCHE FLAVA — MARTIUS ex F.W. SCHULTZ 1829
Orobanche tussilaginis Mutel; *O. froehlichii* Reichenbach fil.

Hellgelbe (Pestwurz-) Sommerwurz

- **ARTBESCHREIBUNG**

Die Pflanzen sind meistens kräftig, selten schlank, sie erreichen eine Größe von etwa 15 bis 65 cm. Der Stengel ist kräftig, aufrecht, orangegelb, gelblichweiß oder bräunlich gefärbt, reichlich mit Drüsenhaaren besetzt, unten reichlich und dicht, oben meist spärlicher beschuppt. Die unteren Schuppen sind dreieckig, eilänglich bis lanzettlich, aufrecht, fast kahl, die oberen Schuppen lanzettlich und drüsenhaarig. Der Blütenstand ist meist dicht- und reichblütig, zylindrisch, später gestreckt und lockerblütiger. Vorblätter sind nicht vorhanden. Das Tragblatt ist lanzettlich, spitz, etwa so lang oder etwas länger als die Blütenkrone, in der Mitte oft etwas abwärts gekrümmt, anfangs rötlich, später schwarzbraun gefärbt, reichlich drüsenhaarig. Die Kelchhälften sind ungespalten oder sehr ungleich zweizähnig, lang zugespitzt, nervig, drüsenhaarig, meist kürzer oder halb so lang und dunkler gefärbt als die Blütenkrone. Die Blüten sind ziemlich groß, zuerst aufrecht bis abstehend, später oft waagrecht-abstehend. Die Blütenkrone ist meistens 15 bis 25 mm lang, röhrig, über der Ansatzstelle der Staubblätter bauchig erweitert, mit hellen Drüsenhaaren besetzt, (hell)gelb bis gelblichweiß oder rötlichbraun (oft beim Blütebeginn schwefelgelb, später rötlichbraun), an der Oberlippe rötlich gefärbt. Die Rückenlinie der Blütenkrone ist vom Grund an gleichmäßig gekrümmt, in der Mitte öfter etwas stärker vorwärts gebogen und manchmal kurz vor der Oberlippe noch etwas aufgerichtet. Die Oberlippe der Blütenkrone ist tief ausgerandet, zweilappig mit zuerst vorgestreckten und später zurückgeschlagenen Lappen und reichlich behaart. Der Mittelzipfel der Unterlippe der Blütenkrone ist meist größer als die Seitenzipfel; am Rande ungleich gezähnelt und spärlich mit Drüsenhaaren besetzt. Die Staubblätter sind 4 bis 6 mm hoch über dem Grund der Kronröhre eingefügt. Die Staubfäden sind von der Basis bis zur Mitte stark behaart und bis zu den Staubbeuteln spärlicher mit Drüsenhaaren besetzt. Die Staubbeutel sind an der Naht behaart. Der Griffel ist spärlich drüsenhaarig oder kahl. Die Narbe besteht aus zwei abgerundeten, kugeligen Lappen und ist orange oder wachsgelb gefärbt, nach der Blüte eingerollt. 2n = 38.

- **BLÜTEZEIT**

Mitte Juni bis Mitte August, je nach Höhenlage.

- **STANDORT**

Die meisten Standorte befinden sich an feuchten Stellen in Steinschutt, Schuttfluren und in Schotterfluren, aber auch an moosigen Stellen in Kiefernwäldern und an Böschungen, auf kalkreichen Kies- und Steinböden. An vielen Standorten wachsen oft viele Pflanzen (gesellig) in großen Gruppen zusammen.

- **WIRT**

Schmarotzt auf *Asteraceae*-Arten; hauptsächlich auf *Petasites paradoxus* (seltener auf anderen *Petasites*-Arten) und auf *Tussilago farfara*. Auch auf *Adenostyles*-Arten.

- **GESAMTVERBREITUNG**

Mittel- und Osteuropa, dort von den französischen Alpen (Jura), die Schweiz, Südostdeutschland, Österreich, der Tschechischen Republik und die Slowakei (Karpaten), Südpolen, Ungarn, Rumänien, ehemaliges Jugoslawien und Norditalien. Wahrscheinlich auch in den Pyrenäen, wurde aber nicht von Saule (1991) bestätigt.
Orobanche flava ist nicht häufig; sie wächst an wenigen Stellen, vor allem im Alpenvorland und in den Flußtälern sind mehrere Standorte bekannt.

- **BEMERKUNGEN**

Die Art ist leicht an ihrem Wirt zu erkennen. Manchmal aber wächst *Orobanche flava* aber auch zwischen *Hedera helix* oder *Salvia glutinosa*. Außerdem findet man sie vor allem im Alpenvorland oder im Mittelgebirge.

Butterbur Broomrape

- **SPECIES DESCRIPTION**

The plant is usually stout, rarely slender, reaching approximately 15-65 cm. The stem is stout, erect, orange-yellow, yellowish-white or brownish, richly glandular-pubescent, densely scaled with numerous scale leaves below and mostly sparsely scaled above. The lower scale leaves are triangular, oval-elongated to lanceolate, erect, almost glabrous; the upper scale leaves are lanceolate and glandular-pubescent. The inflorescence usually has numerous flowers in a dense, cylindrical spike, which becomes elongated and lax later. Bracteoles are absent. The bract is lanceolate, acute, about as long as or slightly longer than the corolla, often slightly deflexed in the middle, reddish in the beginning, brown to black later, richly glandular-pubescent. The calyx-segments are entire or unequally bidentate, elongated-acute, veined, glandular-pubescent, usually shorter than or half as long as the corolla and more darkly coloured than the latter. The flowers are quite large, erect and spreading at first, often horizontal later. The corolla is usually 15-25 mm long, tubular, inflated above the insertion of the stamens, with light glandular hairs, (bright) yellow to yellowish-white or reddish-brown (often sulphur-yellow at the beginning of the flowering, later reddish-brown), reddish at the upper lip. The dorsal line of the corolla is evenly curved from the base onwards, often more clearly bent forward in the middle and sometimes raised a little near the upper lip. The upper lip of the corolla is deeply emarginate, bilobate, richly pubescent with lobes porrect first and deflexed later. The middle lobe of the lower lip is usually larger than the side lobes; it is unevenly crenate and sparsely glandular-pubescent at the margin. The stamens are inserted 4-6 mm above the base of the corolla-tube. The filaments are densely pubescent from the base to the middle and more sparsely pubescent up to the anthers. The anthers are pubescent at the line of fusion. The style is sparsely glandular-pubescent or glabrous. The stigma consists of two rounded, spherical lobes and is orange or wax-coloured, rolled up after flowering. 2n = 38.

- **FLOWERING TIME**

Mid-June to mid-August, depending on altitude.

- **HABITAT**

Most locations are in humid places on stony ground or rubble, but also in mossy places in pine forests and on slopes, on calcareous, stony or gravelly ground. In many locations plants grow together in large groups.

- **HOST**

Parasitic on *Asteraceae* species; mainly on *Petasites paradoxus* (rarely on other *Petasites* species) and on *Tussilago farfara*. Also on *Adenostyles* species.

- **DISTRIBUTION**

Central and eastern Europe, from the French alps (Jura), Switzerland, south-eastern Germany, Austria, the Czech Republic, Slovakia (Carpathian Mountains), southern Poland, Hungary, Rumania, former Yugoslavia and northern Italy. Probably in the Pyrenees too, but not confirmed by Saule (1991).
Orobanche flava is not frequent; it only grows in a few locations, e.g. in the alpine foothills and in river valleys.

- **COMMENTS**

Orobanche flava is easily recognized by its host, although it is occasionally found amidst *Hedera helix* or *Salvia glutinosa*. It is furthermore found mostly in alpine foothills and sub-alpine mountains.

OROBANCHE FLAVA

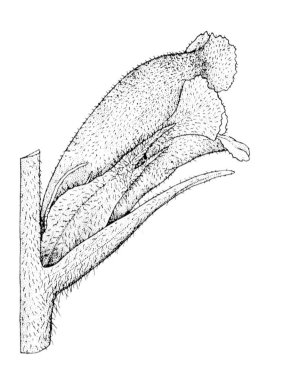
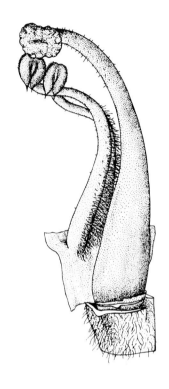
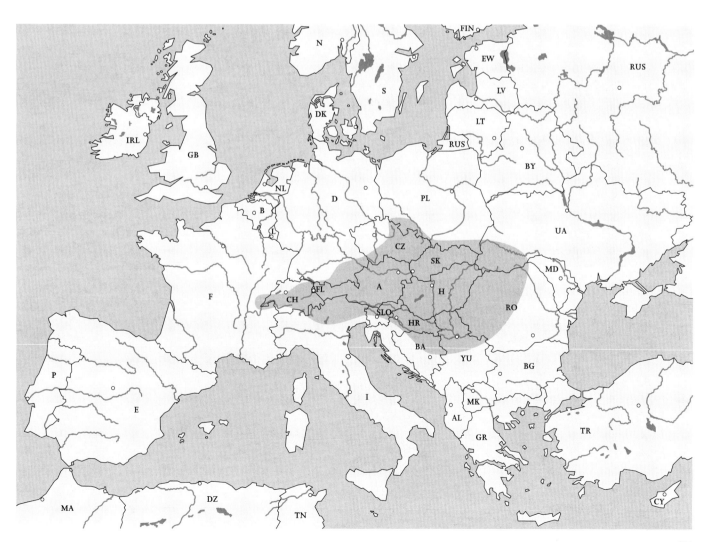

Oberstdorf, Allgäu (D), 2-8-1992

mit Wirtspflanze / with host (*Petasites paradoxus*), Lunz am See, Ybbtaler Alpen (A), 9-7-1994

mit Wirtspflanze / with host (*Petasites paradoxus*), Mundraching, Lechtal (D), 29-6-1992

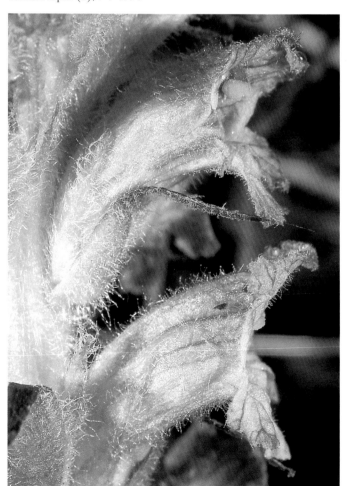

Berchtesgaden, Berchtesgadener Alpen (D), 28-7-1993

Neuberg an der Mürz, Alpen Rax Schneeberg (A), 9-7-1994

4.18

OROBANCHE GRACILIS — SMITH 1798
Orobanche cruenta Bertoloni

Zierliche (Blutrote) Sommerwurz

- **ARTBESCHREIBUNG**

Meistens kräftige Pflanzen, bis zu 60 cm hoch. Der Stengel ist meistens kräftig, selten schlank, aufrecht, hell- oder dunkelgelb bis rötlich oder dunkelrot, selten dunkelbraun gefärbt, unten dicht, oben lockerer beschuppt, zur Gänze reichlich mit Drüsenhaaren besetzt. Die Schuppen sind breit lanzettlich, aufrecht bis abstehend und drüsenhaarig. Der Blütenstand ist meist lockerblütig, langgestreckt mit vielen Blüten, die meistens über den größten Teil des Stengels verteilt sind. Vorblätter sind nicht vorhanden. Das Tragblatt ist etwa so lang wie die Blütenkrone, reichlich drüsenhaarig, breitlanzettlich (an der Basis eiförmig) und abwärts gebogen, rötlichbraun oder schwarzbraun gefärbt. Die Kelchhälften sind mehr oder weniger tief ungleich zweizähnig, selten vorn miteinander verbunden, etwa halb so lang wie die Blütenkrone, reichlich mit Drüsenhaaren besetzt und an der Spitze meistens etwas dunkler gefärbt als die Blütenkrone. Die Blüten sind mittelgroß und abstehend bis aufrecht. Die Blütenkrone ist 11 bis 33 mm lang, röhrig-glockig und über der Ansatzstelle der Staubblätter stark bauchig erweitert, außen mit hellen, kurzen Drüsenhaaren besetzt, innen kahl und glänzend, außen an der Basis hell- bis dunkelgelb gefärbt, darüber rot überlaufen oder zur Gänze rötlich mit dunkelroten Nerven, innen leuchtend dunkelblutrot, selten einfarbig gelb gefärbt. Die Rückenlinie der Blütenkrone ist vom Grund an gleichmäßig gebogen. Die Oberlippe der Blütenkrone ist meistens ungeteilt oder faltig-ausgerandet, mit erst vorgestreckten, später zurückgerollten Lappen. Die Unterlippe der Blütenkrone ist etwas herabgeschlagen mit drei fast gleichgroßen (Mittellappen manchmal etwas größer), abgerundeten, fast aufgerichteten, längsfaltigen Seitenlappen. Die Staubblätter sind am oder nahe dem Grund (bis 2 mm hoch) der Kronröhre eingefügt und dort von einer kleinen Nektardrüse umgeben. Die Staubfäden sind unten mehr oder weniger behaart und oben bis zu den Staubbeuteln mit wenigen Drüsenhaaren besetzt. Die Staubbeutel sind an der Naht spärlich behaart. Der rötliche Griffel ist in der oberen Hälfte mit Drüsenhaaren besetzt. Die Narbe besteht aus zwei abgerundeten, kugeligen Lappen und ist im unteren Teil dunkelgelb und im oberen Teil rötlich gefärbt oder braunrot oder purpurn gerandet. Die Art duftet schwach nach Nelken. 2n = 73-91, 112-116.

- **BLÜTEZEIT**

Juni, manchmal ab Ende Mai. In den höheren Lagen bis September, in wärmeren Lagen auch schon im April und Mai.

- **STANDORT**

Die meisten Standorte befinden sich auf Halbtrockenrasen und Magerwiesen, aber auch an Wald- und Straßenrändern, seltener im Gebüsch auf kalkreichen Lehm- oder Lößböden.

- **WIRT**

Schmarotzt auf *Fabaceae* (vor allem auf *Genista*-, *Lotus*-, *Hippocrepis*-, *Tetragonolobus*-, *Onobrychis*-, *Cytisus*-, *Trifolium*- und *Dorycnium*-Arten).

- **GESAMTVERBREITUNG**

Vor allem in den wärmeren Teilen Mittel- und Südeuropas bis zum Schwarzen Meer. Nördlich bis Nordfrankreich, Zentraldeutschland, Österreich, die Slowakei (auch in Südostpolen), Ungarn und Rumänien. Auch in Transkaukasien (wo sie selten ist) und in Nordafrika. Die Angaben von Antalya und Südwestanatolien (die Türkei) sollten überprüft werden.
Die Art ist weit verbreitet in den Voralpen und Alpen, ziemlich häufig in den niedrigen Lagen von Österreich, stellenweise nicht selten.

- **BEMERKUNGEN**

Obwohl *Orobanche gracilis* sehr variabel ist (Blütenstand, Form, Größe und Farbe der Blumenkrone) ist sie an der Farbe der Blumenkrone, innen leuchtend blutrot und außen meistens hell- bis dunkelgelb, und an der gelben Narbe, die einen roten Rand hat, leicht zu erkennen.

Slender Broomrape

- **SPECIES DESCRIPTION**

The plant is usually stout, up to 60 cm tall. The stem is usually stout, rarely slender, erect, bright yellow or dark yellow to reddish or dark red, rarely dark brown, glandular-pubescent all over, with numerous scale leaves below, more laxly scaled above. The scale leaves are broadly lanceolate, erect to spreading and glandular-pubescent. The inflorescence is usually lax, elongated, with numerous flowers, covering the larger part of the stem. Bracteoles are absent. The bract is approximately half the size of the corolla, richly glandular-pubescent, broadly lanceolate (oval at the base) and deflexed, reddish-brown or brown to black. The calyx-segments are more or less deeply, unequally bidentate, rarely fused at the front, approximately half as long as the corolla, richly glandular-pubescent, the tips usually more darkly tinted than the corolla. The flowers are of medium size, spreading to erect. The corolla is 11-33 mm long, tubular-campanulate, distinctly inflated above the insertion of the stamens, glandular-pubescent with short, light hairs on the outside, glabrous and shiny on the inside; the outside is light to dark yellow at the base, tinged with red or entirely reddish with dark red veins, the inside is intensely dark blood red, rarely simply entirely yellow. The dorsal line of the corolla is evenly curved from the base. The upper lip of the corolla is usually entire or plicate-emarginate, with lobes porrect at first and recurved later. The lower lip of the corolla is slightly deflexed and has three rounded, nearly erect, plicate lobes of nearly equal sizes (the middle sometimes slightly larger). The stamens are inserted at or near the base (up to 2 mm high) of the corolla-tube and surrounded by a small nectar gland. The filaments are more or less pubescent below and sparsely glandular-pubescent above, up to the anthers. The anthers are sparsely pubescent at the line of fusion. The style is reddish and glandular pubescent above. The stigma consists of two rounded, spherical lobes and is dark yellow below and reddish or with a brown-red or purple margin above. The plant emits a weak scent of carnations. 2n = 73-91, 112-116.

- **FLOWERING TIME**

June, sometimes from end of May onwards. In higher altitudes to September, in warmer situations sometimes as early as April or May.

- **HABITAT**

Most locations are in semi-arid grassland and nutrient-poor meadows, also on the edges of forests and on road-sides, more rarely in thickets on calcareous loamy soil or loess.

- **HOST**

Parasitic on *Fabaceae* (especially on *Genista*, *Lotus*, *Hippocrepis*, *Tetragonolobus*, *Onobrychis*, *Cytisus*, *Trifolium* and *Dorycnium* species).

- **DISTRIBUTION**

Mainly in the warmer areas of central and southern Europe to the Black Sea. Northward to northern France, central Germany, Austria, the Slovak Republic (also in south-eastern Poland), Hungary and Rumania. Also in the Transcaucasian region (where it is rare) and in northern Africa. Findings in Antalya and south-western Anatolia (Turkey) should be verified.
Orobanche gracilis grows in a wide area of the sub-alpine regions and of the alps, is quite frequent in low altitude regions of Austria, not rare in some places.

- **COMMENTS**

Although *Orobanche gracilis* is highly variable in its characteristics (inflorescence and shape, size and colour of the corolla), it is easily recognized by the colour of the corolla, intensely blood red on the inside and usually light to dark yellow on the outside, and by its yellow stigma with its red margin.

OROBANCHE GRACILIS

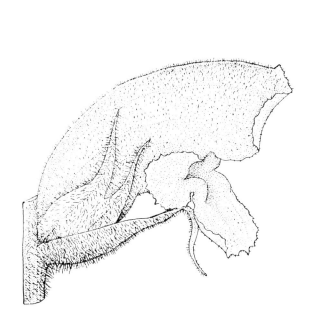
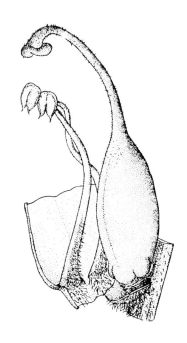

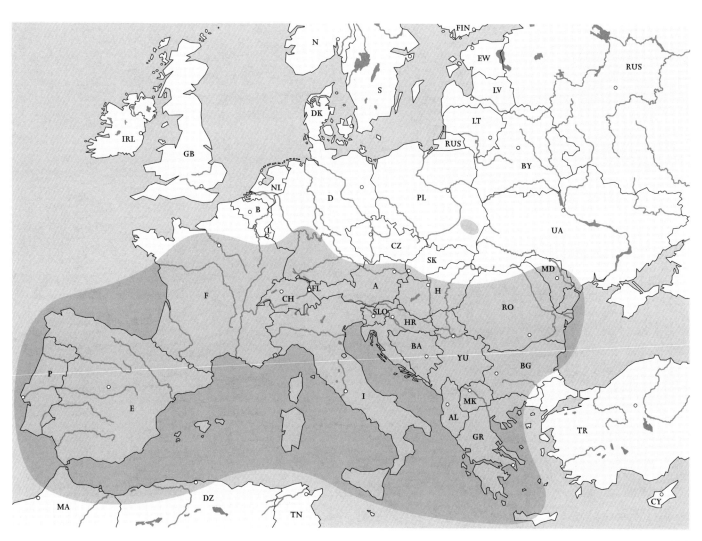

OROBANCHE GRACILIS

Berchtesgaden, Berchtesgadener Alpen (D), 26-7-1989

OROBANCHE GRACILIS

mit Wirtspflanze / with host (*Lotus corniculatus*), Berchtesgaden, Berchtesgadener Alpen (D), 26-7-1989

Les Nouillers, Charente (F), 15-6-1986

Les Nouillers, Charente (F), 15-6-1986

Andrijevica, Montenegro (YU), 11-7-1987

OROBANCHE HEDERAE DUBY 1828

Orobanche medicaginis Reichenbach; *O. laurina* Reichenbach f.

Ivy Broomrape

- **SPECIES DESCRIPTION**

The plants are slender to stout, ca. 10-55 cm tall. The stem is slender, sometimes stout, erect and yellowish, yellow-brown, brown or reddish, sometimes reddish-brown below and yellowish above, seldom yellow or pale over its entire length, with numerous, dense and quite long glandular hairs, sparsely or densely scaled up to the inflorescence. The lower scale leaves are oval to lanceolate and glabrous, upper scale leaves are rather broad or lanceolate and glandular, erect. The spike is lax and many-flowered (at first cylindrical and dense above), later lax below, with the lower flowers set far apart and very low on the stem. Bracteoles are absent. The bract is approximately as long as or longer than the corolla, red-brown to brown or black, densely glandular, lanceolate, with brown to black apex, in the upper parts with deflexed or recurved apex. The calyx-segments are entire, bifid or unequally bidentate, oval at the base and narrowly lanceolate to filiform above, densely glandular, about two thirds or half as long as the corolla, dark red or dark brown, at the base usually lighter than the corolla (brown on light yellow plants). The flowers are of medium size, erect at first, spreading or bent forward later. The corolla is approximately 10-22 mm long, narrowly tubular, inflated above the insertion of the stamens, contracted near the lower lip and widening into a funnel near the margin, with bright glandular hairs, light yellow, yellowish, rarely yellowish-brown. The back of the corolla is tinged with violet or red, especially near the upper lip. The whole of the corolla shows red-violet veins (especially near the upper lip). The dorsal line of the corolla is evenly curved from the base, often almost straight in the middle and sometimes raised near the upper lip. The upper lip of the corolla is entire or emarginate, with usually porrect or spreading lobes, with bright glandular hairs near the tips. The lower lip of the corolla is usually not deflexed and has three dentate, rounded and plicate lobes. The stamens are inserted 3-4 mm above the base of the corolla-tube. The filaments are pubescent below and glabrous above, up to the anthers. The anthers are brown, almost glabrous, and protrude from the flower. The style has sparse, short glandular hairs or is glabrous. The stigma consists of two elongated, globular lobes and is golden yellow, dark yellow, light yellow, orange or reddish yellow (lighter above), dark brown later. 2n = 38.

- **FLOWERING TIME**

May to August, later at high altitudes.

- **HABITAT**

Most plants grow in moist, open, mixed deciduous forests, under shrubs, on roadsides, in shady verges of forests and parks (where it has usually been introduced), in loose, fertile, loamy soil in sheltered locations.

- **HOST**

Parasitic mainly on *Hedera helix*, sometimes on *Pelargonium* and other ornamental plants.

- **DISTRIBUTION**

Mainly in southern Europe (especially in the Mediterranean region) to Asia Minor (Turkey), in the Caucasus countries and Iran. Also in western and central Europe: north to Ireland, southern Scotland, the Netherlands, Germany and Austria. Also in north-western Africa and on the Canary Islands. *Orobanche hederae* has a wide distribution, but is rare everywhere. It is often naturalized in botanical gardens.

- **COMMENTS**

Orobanche hederae grows mainly in moist, shady locations in mixed forests. The host is almost exclusively *Hedera helix*, a species which is easy to identify and covers large areas in many forests. The species has a very lax inflorescence.

OROBANCHE HEDERAE

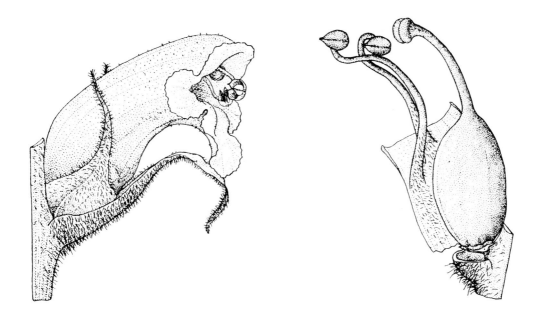

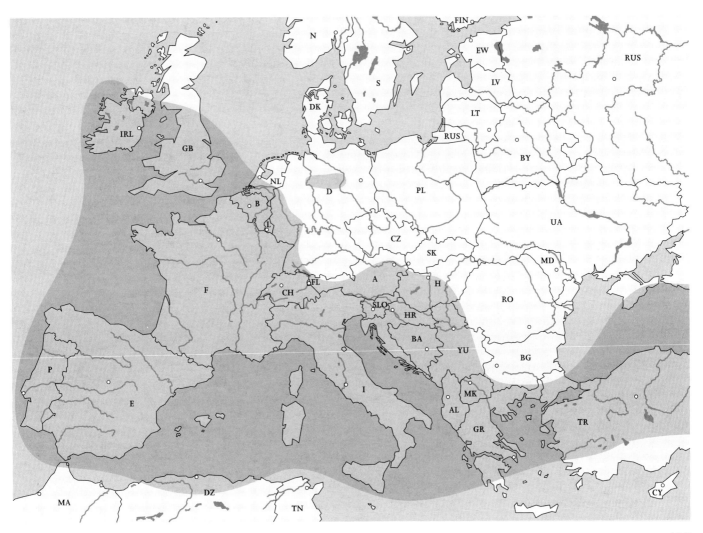

Maastricht, St. Pietersberg (NL), 3-7-1988

mit Wirtspflanze / with host (*Hedera helix*), Maastricht, St. Pietersberg (NL), 3-7-1988

Sorlana, Liguria (I), 4-5-1990

Riva, Lago di Garda (I), 11-6-1980

Macka, Trabzon (TR), 20-5-1988

OROBANCHE LASERPITII-SILERIS   REUTER ex JORDAN 1846

Laserkraut- (Bergkümmel-) Sommerwurz

- ARTBESCHREIBUNG

Die Pflanzen sind meistens kräftig, selten schlank, etwa 30 bis 70 (80) cm hoch. Der Stengel ist kräftig, aufrecht, gelb, gelblichbraun, orangegelb, bräunlich oder rötlich gefärbt, zur Gänze reichlich mit kurzen Drüsenhaaren besetzt, unten reichlich und dicht (dachig), oben meist etwas spärlicher beschuppt. Die unteren Schuppen sind dreieckig und fast kahl, die oberen Schuppen lanzettlich, aufrecht bis abstehend, drüsenhaarig. Der Blütenstand ist zylindrisch, meist dicht- und sehr reichblütig, selten im unteren Teil gestreckt und lockerblütiger; der Blütenstand ist etwa ein Drittel so lang wie der Stengel. Vorblätter sind nicht vorhanden. Das Tragblatt ist lanzettlich, spitz, etwa gleich lang oder wenig länger als die Blütenkrone, spärlich drüsenhaarig oder kahl, in der Mitte abwärts gekrümmt, dunkelbraun gefärbt. Kelch aus zweiteiligen ungleich-zweizähnigen, eiförmigen (seltener ungeteilten), freien oder auf der unteren Seite miteinander verwachsenen Hälften bestehend, drüsenhaarig, meist kürzer oder halb so lang wie die Blütenkrone und dunkler (rotbraun) gefärbt. Die Blüten sind groß, bei beginnender Blüte aufrecht, später abstehend. Die Blütenkrone ist 20 bis 30 mm lang, weitröhrig, über der Ansatzstelle der Staubblätter bauchig erweitert, sehr reichlich mit hellen, wachsgelben Drüsenhaaren besetzt, außen meistens gelblichbraun, rötlichbraun, braunviolett (vor allem gegen den Rand) gefärbt, am Grund und im Bereich der Unterlippe gelblich gefärbt, selten zur Gänze gelblich, mit dunkelvioletten Nerven, innen hellgelb bis gelblich gefärbt. Die Rückenlinie der Blütenkrone ist vom Grund an gleichmäßig gekrümmt. Die Oberlippe der Blütenkrone ist tief zweilappig mit zuerst vorgestreckten und später aufrecht-abstehenden Lappen und spärlich (fast kahl) mit Drüsenhaaren besetzt. Der Mittelzipfel der Unterlippe der Blütenkrone ist größer als die Seitenzipfel; alle drei Lappen sind abgerundet, am Rande gezähnelt und spärlich (fast kahl) mit Drüsenhaaren besetzt. Die Staubblätter sind 5 bis 7 mm hoch über dem Grund der Kronröhre eingefügt, am Grund von einem goldgelben Fleck umgeben. Die Staubfäden sind von der Basis bis zur Mitte stark behaart und bis zu den Staubbeuteln spärlicher mit Drüsenhaaren besetzt. Die Staubbeutel sind fast kahl. Der Griffel ist vor allem im oberen Teil drüsenhaarig. Die Narbe besteht aus zwei abgerundeten, kugeligen Lappen und ist wachsgelb, gelb oder orange gefärbt. 2n = 38.

- BLÜTEZEIT

Mitte Juni bis Anfang August, je nach Höhenlage.

- STANDORT

Die meisten Standorte befinden sich an sonnigen, warmen Gras- und Staudenhalden, in Steinschutt, Schuttfluren und in Schotterfluren auf kalkhaltigen Böden. *Orobanche laserpitii-sileris* ist eine mittel- und südeuropäische Gebirgspflanze.

- WIRT

Schmarotzt auf *Laserpitium siler*, seltener auf *L. latifolium* und *L. halleri*.

- GESAMTVERBREITUNG

Zentraleuropa; vom südlichen Juragebirge und Savoyen (Frankreich) über die Alpen (der Schweiz, Österreich und Norditalien), den nordwestlichen Teil des Balkans (südwärts bis Mazedonien) bis nach Bulgarien. Auch in den Pyrenäen (Pyrenées aragonaises).
Die Art ist sehr selten und fehlt in großen Gebieten. Meistens tritt sie nur zerstreut auf.

- BEMERKUNGEN

Die Art wächst nur in den höheren Gebirgslagen. Die Pflanzen sind sehr kräftig und könnten mit *Orobanche rapum-genistae* verwechselt werden.

Laserpitium Broomrape

- SPECIES DESCRIPTION

The plant is usually stout, rarely slender, approximately 30-70 (80) cm tall. The stem is stout, erect, yellow, yellowish-brown, orange-yellow, brownish or reddish, richly glandular-pubescent all over, with short glandular hairs, richly and densely scaled (imbricate) below, usually more laxly scaled above. The lower scale leaves are triangular and almost glabrous, the higher scale leaves are lanceolate, erect to spreading and glandular-pubescent. The inflorescence usually has numerous flowers in a dense, cylindrical spike, rarely elongated and more lax in the lower part; the inflorescence covers about one third of the stem. Bracteoles are absent. The bract is lanceolate, acute, about as long as or slightly longer than the corolla, sparsely glandular-pubescent or glabrous, deflexed in the middle, dark brown in colour. The calyx consists of bifid, unequally bidentate, oval (rarely entire) halves, which are free or fused below, glandular-pubescent, usually shorter than or about half as long as the corolla; it has a darker colour (red-brown) than the latter. The flowers are large, erect at the beginning of flowering, spreading later. The corolla is 20-30 mm long, widely tubular, inflated above the insertion of the stamens, very richly glandular-pubescent with light, wax-coloured glandular hairs; on the outside usually yellowish-brown, reddish-brown, brownish-violet (especially towards the margin); yellowish at the base and near the lower lip, rarely yellow all over, with dark violet veins; bright yellow to yellowish on the inside. The dorsal line of the corolla is evenly curved from the base. The upper lip of the corolla is deeply bilobate, its lobes porrect at first and erect to spreading later; sparsely glandular-pubescent (almost glabrous). The central lobe of the lower lip of the corolla is larger than the side lobes; all three lobes are rounded, crenate at the margin and sparsely glandular-pubescent (almost glabrous). The stamens are inserted 5-7 mm above the base of the corolla-tube, their base enclosed by a golden yellow spot. The filaments are richly pubescent from the base up to the middle and sparsely glandular-pubescent above, up to the anthers. The anthers are almost glabrous. The style is glandular-pubescent especially in the upper part. The stigma consists of two rounded, spherical lobes and is wax-coloured, yellow or orange. 2n = 38.

- FLOWERING TIME

Mid-June to beginning of August, depending on altitude.

- HABITAT

Most locations are in sunny, warm, grassland or herbaceous vegetations, on rubble or stony or gravelly ground, on calcareous soil. *Orobanche laserpitii-sileris* is a plant of the central and southern European mountains.

- HOST

Parasitic on *Laserpitium siler*, more rarely on *L. latifolium* and *L. halleri*.

- DISTRIBUTION

Central Europe; from the southern Jura mountains and Savoy (France) through the Alps (Switzerland, Austria and northern Italy), the northwestern part of the Balkans (southward to Macedonia) to Bulgaria. Also in the Pyrenees (Pyrenées argonaises).
The plant is very rare and is absent from large areas. Its distribution is highly sporadic.

- COMMENTS

Orobanche laserpitii-sileris only grows in higher altitudes. The plant is very stout and could be mistaken for *Orobanche rapum-genistae*.

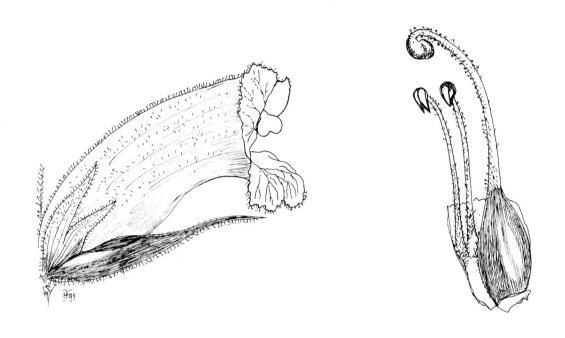
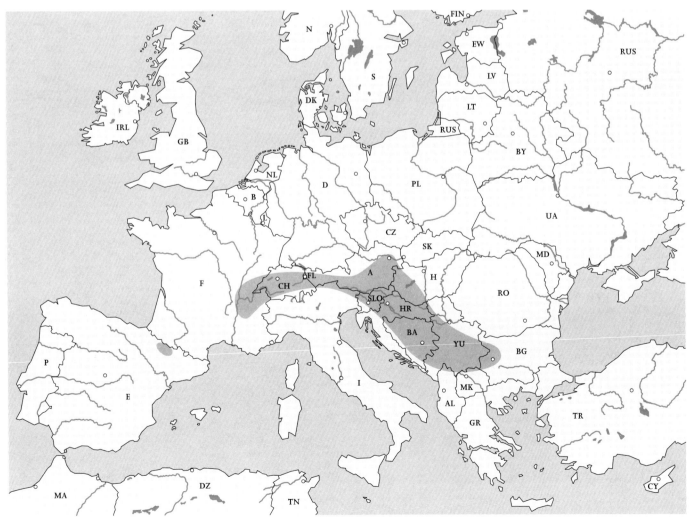

Le Flon, Lac de Taney (CH), 4-7-1992

mit Wirtspflanze / with host (*Laserpitium siler*), Le Flon, Lac de Taney (CH), 4-7-1992

mit Wirtspflanze / with host (*Laserpitium siler*), Le Flon, Lac de Taney (CH), 4-7-1992

Le Flon, Lac de Taney (CH), 4-7-1992

Le Flon, Lac de Taney (CH), 4-7-1992

4.21

OROBANCHE LUCORUM
Orobanche berberidis Facchini

A. BRAUN ex KOCH 1833

Hain- (Berberitzen-) Sommerwurz

- **ARTBESCHREIBUNG**

Pflanzen meist kräftig, etwa 10 bis 60 cm hoch. Der Stengel ist schlank bis kräftig, aufrecht, meistens (gold)gelb, gelbbraun oder rosa gefärbt, drüsenhaarig, bis zum Blütenstand unten reichlich, oben spärlicher beschuppt. Die unteren Schuppen sind dreieckig und fast kahl, die oberen dreieckig bis lanzettlich und sind mit Drüsenhaaren besetzt, aufrecht bis abstehend. Der Blütenstand ist meist dicht- und reichblütig, zylindrisch, später lockerblütig und gestreckt (verlängert). Vorblätter sind nicht vorhanden. Das Tragblatt ist etwa so lang wie die Blütenkrone, dunkelbraun gefärbt, spärlich mit Drüsenhaaren besetzt oder kahl, lanzettlich mit schwarzbrauner, herabgeschlagener Spitze, oft waagrecht abstehend. Die Kelchhälften sind ungespalten oder ungleich zweizähnig, meist nur halb so lang wie die Blütenkrone, drüsenhaarig, genervt und von gleicher Farbe wie die Blütenkrone oder an der Basis heller gefärbt. Die Blüten sind mittelgroß, waagrecht-abstehend oder vorwärts gebogen. Die Blütenkrone ist meistens 12 bis 21 mm lang, über der Ansatzstelle der Staubblätter erweitert mit vielen, hellen Drüsenhaaren besetzt, gelbbraun, rötlichbraun oder orange bis rötlichgelb (selten zur Gänze strohgelb) gefärbt. Die Rückenlinie der Blütenkrone ist von der Basis bis zur Mitte stark nach vorne gekrümmt und von der Mitte bis zur Oberlippe fast gerade oder zur Gänze gleichmäßig gekrümmt. Die Oberlippe der Blütenkrone ist gekielt, ausgerandet oder zweilappig mit vorgestreckten Lappen. Die Unterlippe der Blütenkrone ist herabgeschlagen mit drei gezähnelten, abgerundeten Lappen. Die Staubblätter sind 2 bis 3 mm hoch über dem Grund der Kronröhre eingefügt, am Grund von einer goldgelben Nektardrüse umgeben. Die Staubfäden sind am Grund reichlich behaart und oben bis zu den Staubbeuteln meist spärlich drüsenhaarig oder fast kahl. Die Staubbeutel sind an der Naht behaart. Der Griffel ist selten stark drüsig behaart, meistens kahl oder spärlich mit Drüsenhaaren besetzt. Die Narbe besteht aus zwei goldgelben oder orange, später bräunlich, gefärbten Lappen. Die Blüten duften schwach. $2n = 38$.

- **BLÜTEZEIT**

Ende Juni bis August, manchmal noch im September.

- **STANDORT**

Die meisten Standorte befinden sich in Auengebüschen, in Gebüschsäumen, aber auch in frischen, schwach gedüngten Wiesen an schattigen Stellen auf basenreichen, feuchten, steinig-kiesigen Ton- und Lehmböden. An vielen Standorte wachsen oft viele Pflanzen in großen Gruppen zusammen. Vielfach in botanischen Gärten eingebürgert, da sich *Orobanche lucorum* leicht auf *Berberis*-Wurzeln kultivieren läßt (nördlich bis Litauen, Botanischer Garten von Vilnius).

- **WIRT**

Schmarotzt vor allem auf *Berberis vulgaris*, seltener auf *Rubus*- oder *Crataegus*-Arten.

- **GESAMTVERBREITUNG**

Mitteleuropa, dort in den Alpen von Süddeutschland (Oberbayern und Oberschwaben), die Schweiz (Tessin und Graubünden), Liechtenstein, Österreich (Kärnten, Vorarlberg, Salzburg und Tirol), Italien (Südtirol) und Slowenien. Nach Pignatti (1982) auch in Mittel- und Süditalien, aber diese Angaben, sowie aus Istrien sind irrig (Meusel *et al.*, 1978). Die Angaben von Rumänien (Savulescu *et al.*, 1963) sollten überprüft werden.
Orobanche lucorum bewohnt ein sehr beschränktes Gebiet. In manche Gegenden (Alpenvorland) ist sie nicht selten und tritt häufig auf.

- **BEMERKUNGEN**

Von *Orobanche lucorum* ist die Form 'kirroantha' Beck 1930 beschrieben. Bei dieser Sippe ist die Blumenkrone sowie die ganze Pflanze strohgelb gefärbt. Ihre Blüten sind meistens kleiner, etwa 12 bis 17 mm lang.
Orobanche lucorum wächst meistens in größeren Gruppen zusammen. Ihr Vorkommen unter *Berberis vulgaris* ist ein weiteres, deutliches Unterscheidungsmerkmal.

Barberry Broomrape

- **SPECIES DESCRIPTION**

The plant is usually stout, approximately 10-60 cm tall. The stem is slender to stout, erect, usually (golden) yellow, yellow-brown or pink, glandular-pubescent, densely scaled below and sparsely scaled above, up to the inflorescence. The lower scale leaves are triangular and almost glabrous, the higher ones triangular to lanceolate, glandular-pubescent and erect to spreading. The inflorescence usually has numerous flowers in a dense, cylindrical spike, lax and elongated later. Bracteoles are absent. The bract is about as long as the corolla, dark brown, sparsely glandular-pubescent or glabrous, lanceolate with a brown to black, deflexed tip, often spreading to horizontal. The calyx-segments are entire or unequally bidentate, usually half as long as the corolla, glandular-pubescent, veined, of the same colour as the corolla or more lightly coloured at the base. The flowers are of medium size, horizontal or bent forwards. The corolla is usually 12-21 mm long, inflated above the insertion of the stamens, with numerous, light glandular hairs, yellow-brown, reddish-brown or orange to reddish-yellow (rarely straw-yellow all over). The dorsal line of the corolla is strongly curved forwards from the base to the middle and almost straight from the middle to the upper lip; or evenly curved over its entire length. The upper lip of the corolla is keeled, emarginate or bilobate with porrect lobes. The lower lip of the corolla is deflexed and has three crenate, rounded lobes. The stamens are inserted 2-3 mm above the base of the corolla-tube and surrounded by a golden-yellow nectar gland. The filaments are richly pubescent at the base and usually sparsely glandular-pubescent or almost glabrous above, up to the anthers. The anthers are pubescent at the line of fusion. The style is usually glabrous or sparsely glandular-pubescent, rarely richly glandular-pubescent. The stigma consists of two lobes, which are golden yellow or orange at first and brownish later. The flowers are weakly fragrant. $2n = 38$.

- **FLOWERING TIME**

End of June to August, sometimes as late as September.

- **HABITAT**

Most locations are in marshy woodland, thickets, also on lightly fertile meadows in shaded places on alkaline, moist, clayey and loamy soil with stones and gravel. In many locations numerous plants grow together in large groups. Often naturalized in botanical gardens, as *Orobanche lucorum* is easily cultivated on roots of *Berberis* (northward to Latvia, Vilnius Botanical Gardens).

- **HOST**

Parasitic preferably on *Berberis vulgaris*, more rarely on *Rubus* and *Crataegus* species.

- **DISTRIBUTION**

Central Europe, from the alps of southern Germany (Oberbayern and Oberschwaben), Switzerland (Ticino and Graubünden), Liechtenstein, Austria (Kärnten, Vorarlberg, Salzburg and Tirol), Italy (Alto Adige) and Slovenia. According to Pignatti (1982) also in central and southern Italy, but these identifications are erroneous, as are those of Istria (Meusel *et al.*, 1978). Findings of *Orobanche lucorum* in Rumania (Savulescu *et al.*, 1963) should be verified.
O. lucorum grows in a very limited range. In some areas (alpine foothills) it is not rare.

- **COMMENTS**

The form 'kirroantha' Beck 1930 of *Orobanche lucorum* has been described. This form is yellow all over, including the inflorescence. Its flowers are usually smaller, approximately 12-17 mm long.
Orobanche lucorum usually grows in large groups. Its presence under *Berberis vulgaris* is another clear distinguishing mark.

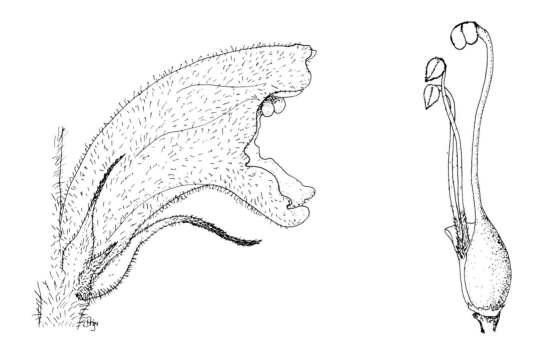

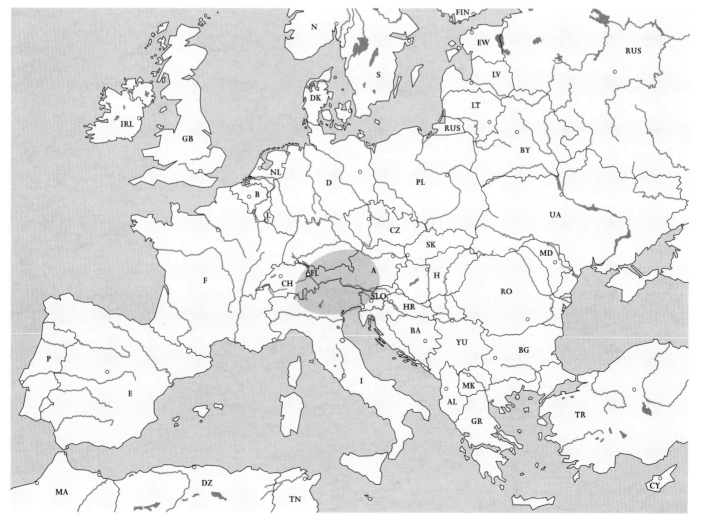

OROBANCHE LUCORUM

mit Wirtspflanze / with host (*Berberis vulgaris*), Oberstdorf, Allgäu (D), 3-8-1992

OROBANCHE LUCORUM

mit Wirtspflanze / with host (*Berberis vulgaris*), Oberstdorf, Allgäu (D), 3-8-1992

Oberstdorf, Allgäu (D), 3-8-1992

Bad Wörishofen, Bayern (D), 2-8-1992

Bad Wörishofen, Bayern (D), 2-8-1992

4.22

OROBANCHE LUTEA — BAUMGARTEN 1816

Orobanche elatior Koch et Ziz, *O. rubens* Wallroth, *O. bueckii* A. Dietrich; *O. medicaginis* Duby

Gelbe (Rötlichgelbe) Sommerwurz

- **ARTBESCHREIBUNG**

Pflanzen von mittelmäßiger Größe, die etwa eine Höhe von 10 bis 60 cm erreichen. Der Stengel ist schlank bis kräftig, aufrecht, meistens gelb, selten braun, rötlichbraun oder purpurn gefärbt, reichlich mit Drüsenhaaren besetzt, bis zum Blütenstand unten reichlich, oben meist spärlicher beschuppt. Die Schuppen sind kurz, länglich-lanzettlich, aufrecht bis abstehend. Der Blütenstand ist meist dicht- und reichblütig, zylindrisch, selten zur Gänze lockerblütig, später im mittleren und unteren Teil des Blütenstandes gestreckt. Vorblätter sind nicht vorhanden. Das Tragblatt ist etwa gleich lang oder etwas kürzer als die Blütenkrone, schwarzbraun (an der Basis gelblich) gefärbt, ab der Mitte abwärts gebogen, helldrüsig. Die Kelchhälften sind fast gleich bis ungleich zwei-zähnig, röhrig-glockig und meist etwa ein Drittel so lang wie die Blütenkrone, genervt und gleich gefärbt wie die Blütenkrone. Die Blüten sind mittelgroß und fast waagrecht-abstehend. Die Blütenkrone ist meistens 20 bis 33 mm lang, über der Ansatzstelle der Staubblätter etwas erweitert mit hellen Drüsenhaaren, hellbraun oder rötlichbraun, selten ganz gelb oder rötlich gefärbt, im unteren Teil bleicher, oft mit intensiver gefärbten Nerven. Die Rückenlinie der Blütenkrone ist schon ab der Basis nach vorne gekrümmt, in der Mitte gerade und im Bereich der Oberlippe leicht oder stark abwärts gebogen. Die Oberlippe der Blütenkrone ist gekielt, ausgerandet oder zweilappig mit aufwärts oder vorgestreckten Lappen. Die Unterlippe der Blütenkrone ist herabgeschlagen mit drei gezähnelten, abgerundeten Lappen. Die Staubblätter sind 3 bis 7 mm hoch über dem Grund der Kronröhre eingefügt, am Grund mit orangegelbem Nektarfleck umgeben. Die Staubfäden sind am Grund reichlich behaart, in der Mitte fast kahl und oben bis zu den Staubbeuteln meist spärlich drüsenhaarig. Die Staubbeutel sind lang wollig behaart. Der Griffel ist vor allem im oberen Teil drüsig behaart. Die Narbe besteht aus zwei auffällig, kräftig gelben, selten orange gefärbten Lappen. Blüten meist geruchlos. $2n = 38$.

- **BLÜTEZEIT**

Ende Mai bis Juli. Eine relativ frühblühende Art.

- **STANDORT**

Viele Standorte befinden sich auf Trocken- und Halbtrockenrasen an sonnigen Stellen, aber auch in grasigen Hängen, seltener in Wiesen und Luzernenfeldern auf basenreichen Lehm-, Sand- oder Lößböden.

- **WIRT**

Schmarotzt auf *Fabaceae* (besonders auf *Medicago-*, *Trifolium-* und *Melilotus-*, seltener auf *Lotus-* und *Dorycnium-*Arten). Auf Kulturpflanzen in Luzerne- oder Kleefeldern.

- **GESAMTVERBREITUNG**

Mitteleuropa (nördlich bis in die Niederlande, Norddeutschland, Polen und die baltische Staaten) und nördliches Südeuropa (von Nordostspanien (Katalonien), über Italien (Sardinien und Sizilien) und Nordgriechenland) bis in die Kaukasusländer, Iran und Mittelasien. Nicht auf Korsika.
Die Art ist in großen Teilen weit verbreitet und häufig, in einigen Gebieten schwerpunktmäßig verbreitet und in anderen Teilen Europas, wie zum Beispiel den Niederlanden, fast ausgestorben. In den Mittelmeerländern (Italien und Balkanhalbinsel) sehr zerstreut.

- **BEMERKUNGEN**

Von *O. lutea* wurde die Varietät '*buekiana*' (Koch) Beck 1890 beschrieben. Von dieser Sippe ist die Blumenkrone stark bogig gekrümmt, fast waagrecht abstehend und bleichgelb oder rötlich gefärbt. Sie schmarotzt vor allem auf *Fabaceae*-Arten.
Orobanche lutea wächst meistens an extrem sonnigen Trockenrasenstandorten auf Sandböden. Die ganze Pflanze ist meistens hell- bis dunkelgelb gefärbt, die Narbe auffallend gelb.

Yellow Broomrape

- **SPECIES DESCRIPTION**

The plant is of medium height, reaching approximately 10-60 cm. The stem is slender to stout, erect, usually yellow, rarely brown, reddish-brown or purple, richly glandular-pubescent, densely scaled below, more sparsely scaled above, up to the inflorescence. The scale leaves are short, elongated-lanceolate, erect to spreading. The inflorescence usually has numerous flowers in a dense, cylindrical spike, rarely lax over its entire length, the middle and lower part of the inflorescence later elongated. Bracteoles are absent. The bract is about as long as or shorter than the corolla, brown to black (yellowish at the base), deflexed from the middle, glandular-pubescent with light hairs. The calyx-segments consist of two equally to unequally bidentate halves, tubular-campanulate, veined, usually about a third of the length of the corolla and of the same colour as the latter. The flowers are medium sized and spreading to almost horizontal. The corolla is usually 20-33 mm long, slightly inflated above the insertion of the stamens, with light glandular hairs, light brown or reddish-brown, rarely entirely yellow or reddish, paler below, often with more intensely coloured veins. The dorsal line of the corolla is curved forward from the base, straight in the middle and more or less deflexed near the upper lip. The upper lip of the corolla is keeled, emarginate, or bilobate, with erect or porrect lobes. The lower lip of the corolla is deflexed and consists of three crenate, rounded lobes. The stamens are inserted 3-7 mm above the base of the corolla-tube, they are surrounded by a orange-yellow nectar spot at the base. The filaments are densely pubescent at the base, almost glabrous in the middle, usually sparsely glabrous-pubescent above, up to the anthers. The anthers are woolly, with long hairs. The style is glandular-pubescent, especially above. The stigma consists of two distinctly yellow, rarely orange, lobes. Flowers usually have no scent. $2n = 38$.

- **FLOWERING TIME**

End of May to July. This species flowers relatively early in the year.

- **HABITAT**

Many locations are in arid and semi-arid grassland in sunny places, also on grassy slopes, more rarely in meadows and lucerne fields on alkaline sand, loamy soil and loess.

- **HOST**

Parasitic mainly on *Fabaceae* (especially on *Medicago*, *Trifolium* and *Melilotus* species, more rarely on *Lotus* and *Dorycnium* species). On cultivated plants in lucerne or clover fields.

- **DISTRIBUTION**

Central Europe (northward to the Netherlands, northern Germany, Poland and the Baltic states) and northern parts of southern Europe (from northeastern Spain (Catalonia), through Italy (Sardinia and Sicily) and northern Greece) to the Caucasian countries, Iran and central Asia. Not on Corsica.
Orobanche lutea is frequent in wide areas, very frequent in some regions, but nearly extinct in other regions, as in the Netherlands. Sporadic in the Mediterranean countries (Italy and the Balkans).

- **COMMENTS**

The variety *Orobanche lutea* var. *buekiana* (Koch) Beck 1890 has been described. The corolla of this variety is strongly curved, spreading to almost horizontal, pale yellow or reddish. It is parasitic mainly on *Fabaceae* species.
Orobanche lutea usually grows in extremely sunny, dry grassland on sandy soil. The entire plant is usually bright yellow to dark yellow, the stigma conspicuously yellow.

OROBANCHE LUTEA

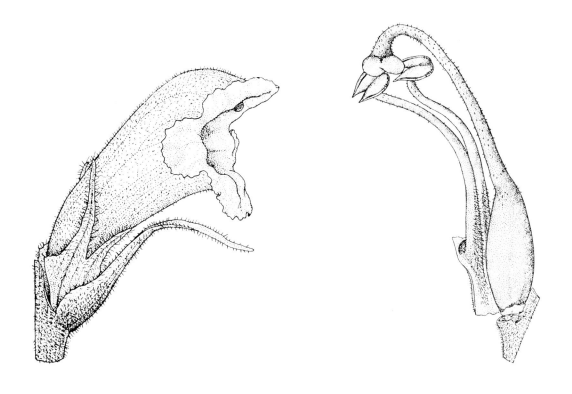

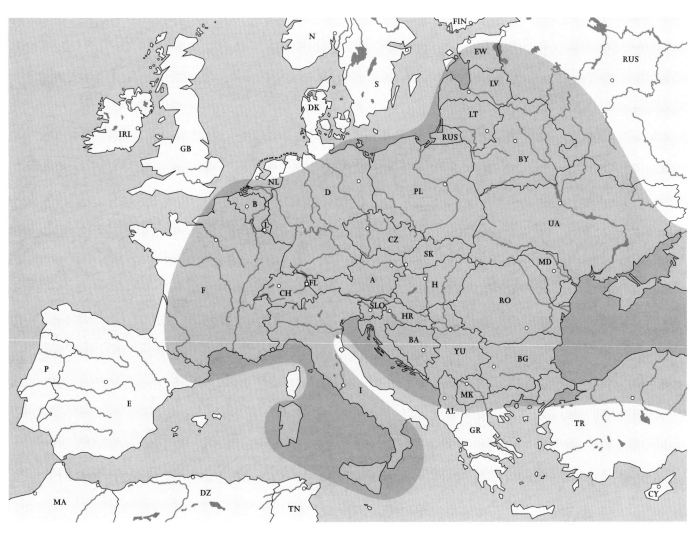

OROBANCHE LUTEA

Suchovske, Bílé Karpaty (CZ), 28-5-1994

OROBANCHE LUTEA

Mainz, Rheinland-Pfalz (D), 20-5-1989

Suchovske, Bílé Karpaty (CZ), 28-5-1994

Tolkamer, Gelderland (NL), 15-6-1985

Gumpoldskirchen, Niederösterreich (A), 29-5-1994

4.23

OROBANCHE MINOR
J.E. SMITH 1797

Orobanche barbata Poiret; *O. nudiflora* Wallroth; *O. apiculata* Wallroth; *O. trifolii pratensis* Schultz

Kleine Sommerwurz

- **ARTBESCHREIBUNG**

Pflanzen schlank, in Kulturgewächsen meistens relativ kräftig, etwa 10 bis 70 cm hoch. Der Stengel ist schlank, manchmal kräftig, aufrecht oder schwach gebogen, meistens rotbraun, violettpurpurn, braunrot oder rötlich überlaufen, selten gelblich, gelblichbraun oder bräunlich gefärbt, dicht drüsenhaarig, im unteren Teil ziemlich dicht, im oberen Teil spärlicher mit Schuppen besetzt. Die unteren Schuppen sind dreieckig-eiförmig, die oberen lanzettlich bis länglich und drüsenhaarig, aufrecht oder abstehend. Der Blütenstand ist anfangs meist zylindrisch, dicht- und reichblütig, später (manchmal mit Ausnahme vom oberen Teil) langgestreckt und lockerblütig (untere Blüten weit voneinander entfernt), wobei die Blüten über etwa drei Viertel der Stengel verteilt sind. Vorblätter sind nicht vorhanden. Das schmallanzettliche Tragblatt ist meistens so lang wie die Blütenkrone, hell- bis dunkelbraun gefärbt, reichlich mit Drüsenhaaren besetzt, und meistens etwas abwärts gebogen. Die Kelchhälften sind ungespalten oder ungleich zweizähnig, meist tief zweispaltig, an der Basis eiförmig und an der Spitze fast fadenförmig, etwa halb so lang wie die Blütenkrone, manchmal auch die Blütenkrone erreichend oder etwas überragend, genervt, drüsenhaarig, meist dunkler (braunrot) als die Blütenkrone gefärbt. Die Blüten sind klein, aufrecht bis abstehend oder vorwärts geneigt. Die Blütenkrone ist etwa 10 bis 19 mm lang, röhrig, gleichmäßig erweitert und gegen den Schlund zu wenig erweitert mit kurzen hellen Drüsenhaaren, gelblich, gelblichweiß bis dunkelgelb, im Bereich der Oberlippe violett gefärbt mit vor allem im Bereich der Oberlippe violettrötlichen Adern. Die Rückenlinie der Blütenkrone ist meistens zur Gänze gleichmäßig gebogen oder im unteren Bereich etwas geknickt, im mittleren und oberen Teil fast gerade und im Bereich der Oberlippe etwas aufgerichtet. Die Oberlippe der Blütenkrone ist fast ungeteilt oder zweispaltig, ausgerandet mit vorwärts gerichteten Lappen. Die Unterlippe der Blütenkrone besteht aus drei fast gleichgroßen, rundlichen, gefalteten, ungleich gezähnelten, herabgebogenen Lappen. Die Staubblätter sind 2 bis 3 mm hoch über dem Grund der Kronröhre eingefügt. Die Staubfäden sind an der Basis spärlich behaart und oben bis zu den Staubbeuteln spärlich mit Drüsenhaaren besetzt oder kahl. Die Staubbeutel sind an der Naht oft behaart. Der Griffel ist spärlich mit Drüsenhaaren besetzt oder kahl. Die Narbe besteht aus zwei halbkugeligen Lappen und ist rosa, braunlila, purpurn, purpurbräunlich, braunviolett, rotviolett, selten weiß oder gelb gefärbt. 2n = 38.

- **BLÜTEZEIT**

Mitte Mai bis Ende August.

- **STANDORT**

Viele Standorte befinden sich auf Fett- und Frischwiesen, auch in ruderal beeinflußten Halbtrockenrasen, in niedrigen, warmen Lagen auf lockeren trockenen, nährstoffreichen Lehm- oder Lößböden. Auch in landwirtschaftlichen Kulturen, wie Kleeäckern, wo sie dann oft die ganze Ernte vernichten kann.

- **WIRT**

Auf *Trifolium*-Arten (*T. pratense*, *T. medium* und *T. arvense*), aber auch auf anderen *Fabaceae*- und *Asteraceae*-Arten schmarotzend. Bei Kulturpflanzen vor allem auf *Trifolium*-Arten, auf *Lotus corniculatus* und auf *Medicago sativa*.

- **GESAMTVERBREITUNG**

Ursprünglich nur in den Mittelmeerländern vorkommend, ist *Orobanche minor* heute weit verbreitet. Durch Einschleppung mit Klee-, Luzerne- und Esparsettesamen (*Trifolium*, *Medicago sativa* und *Onobrychis viciifolia*) hat sie sich bis nach Mittel- und Nordeuropa (nördlich bis Irland, Südschottland, die Niederlande, Dänemark und Südschweden) ausgebreitet. In diesen Ländern tritt *O. minor* überwiegend segetal (vor allem auf *T. pratense*), seltener auch in Frischwiesen auf. Auch in großen Teilen von Afrika, Nordamerika und Neuseeland.
Die Art ist zwar ziemlich verbreitet, aber nicht häufig und unbeständig, oft aber in großen Beständen.

- **BEMERKUNGEN**

Von *Orobanche minor* wurden zwei Varietäten, nämlich *O. minor* var. *maritima* (Pugsley) Rumsey & Jury und *O. minor* var. *compositarum* Pugsley beschrieben. Aus Platzgründen werden diese beiden Varietäten im Kapitel 2.3 (Taxonomie und Nomenklatur) besprochen.
Orobanche minor ist eine sehr variabele Art (siehe seite 158), sie weist eine große Zahl von Formen auf, wodurch eine richtige Bestimmung manchmal schwierig ist. Ein wichtiges Merkmal ist ihre Blütengröße, die meist klein ist. Weiter ist sie an den dunkelvioletten Adern ihrer Blumenkrone zu erkennen.

Common (Small, Lesser) Broomrape

- **SPECIES DESCRIPTION**

The plant is slender, usually quite stout on cultivated plants, approximately 10-70 cm tall. The stem is slender, sometimes stout, erect or slightly curved, usually red-brown, violet-purple, brown-red or tinged with red, rarely yellow, yellow-brown or brownish, densely glandular-pubescent; rather densely scaled below, sparsely above. The lower scale leaves are triangular-oval, the higher ones are lanceolate to elongated, glandular-pubescent and erect or spreading. The inflorescence usually has numerous flowers in a dense, cylindrical spike at first, becoming lax (lower flowers far apart) and elongated later (except for the upper part), with flowers covering three quarters of the stem. Bracteoles are absent. The bract is narrowly lanceolate and usually about as long as the corolla, light to dark brown, richly glandular-pubescent and usually slightly deflexed. The calyx-segments are entire or unequally bidentate, usually deeply bifid, oval at the base and almost filiform at the tip, about half as long as the corolla, sometimes as long as or longer than the latter, veined, glandular-pubescent, usually darker (brown-red) than the corolla. The flowers are small, erect to spreading or bent forwards. The corolla is approximately 10-19 mm long, tubular, gradually widening and slightly inflated near the throat, with light, short glandular hairs, yellowish, yellowish-white to dark yellow, violet or with violet-reddish veins near the upper lip. The dorsal line of the corolla is usually evenly curved over its entire length or slightly geniculate in the lower part, almost straight in the middle and upper part and slightly raised near the upper lip. The upper lip of the corolla is almost entire or bilobate with emarginate, porrect lobes. The lower lip of the corolla consists of three rounded, plicate, unevenly crenate, deflexed lobes of almost equal size. The stamens are inserted 2-3 mm above the base of the corolla-tube. The filaments are sparsely pubescent at the base, sparsely glandular-pubescent up to the anthers. The anthers are often pubescent at the line of fusion. The style is sparsely glandular-pubescent or glabrous. The stigma consists of two hemispherical lobes, which are pink, brown-violet, purple, purple-brown, brown-violet, red-violet, rarely white or yellow. 2n = 38.

- **FLOWERING TIME**

Mid-May to end of August.

- **HABITAT**

Many locations are in fertile meadows, also in ruderal semi-arid grasslands, in warm, low altitude regions, on loose, nutrient-rich loamy soil or loess. Also in agricultural fields, e.g. clover fields, where it can completely destroy the harvest.

- **HOST**

Parasitic on *Trifolium* species (*T. pratense*, *T. medium* and *T. arvense*), as well as on other *Fabaceae* and *Asteraceae* species. Cultivated hosts include *Trifolium* species, as well as *Lotus corniculatus* and *Medicago sativa*.

- **DISTRIBUTION**

Orobanche minor was originally a plant of the Mediterranean countries, but is widely distributed today. Having been imported with seeds of clover, lucerne and sainfoin (*Trifolium*, *Medicago sativa* and *Onobrychis viciifolia*), it has now spread to central and northern Europe (northward to Ireland, southern Scotland, the Netherlands, Denmark and southern Sweden). In all these countries *O. minor* is primarily segetal (mainly on *Trifolium pratense*), more rarely in freshly fertilized meadows. Also in large parts of Africa, North America and New Zealand.
Orobanche minor is widely distributed, but infrequent and fluctuating, often forming large populations.

- **COMMENTS**

Two varieties of *Orobanche minor* have been described: *O. minor* var. *maritima* (Pugsley) Rumsey & Jury and *O. minor* var. *compositarum* Pugsley. For reasons of space, these two varieties have been discussed in chapter 2.3, on Taxonomy and Nomenclature.
Orobanche minor is a highly variable species (see page 158), with a large number of different forms, which may make identification difficult. The size of the flowers (usually small) is an important characteristic, as are the dark violet veins of the corolla.

OROBANCHE MINOR

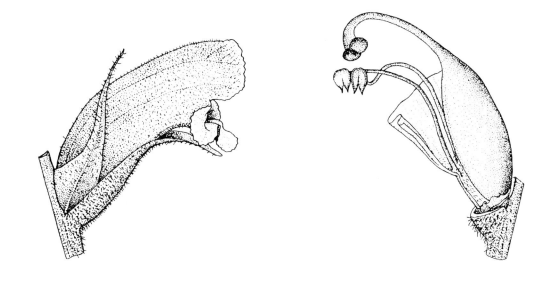

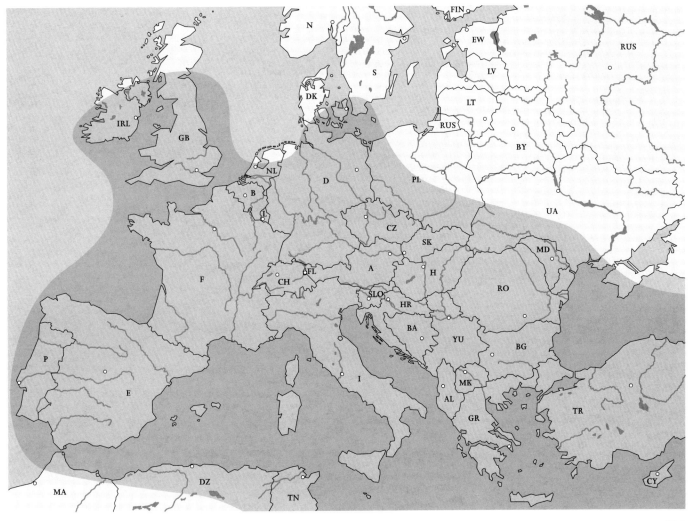

mit Wirtspflanze / with host (*Trifolium pratense*), Landgraaf, Zuid-Limburg (NL), 18-7-1993

mit Wirtspflanze / with host (*Trifolium repens*), Eys, Zuid-Limburg (NL), 5-7-1994

mit Wirtspflanze / with host (*Trifolium pratense*), Landgraaf, Zuid-Limburg (NL), 18-7-1993

Bonifacio, Corse (F), 17-4-1990

Siniscola, Sardagena (I), 27-4-1990

4.24

OROBANCHE PALLIDIFLORA — WIMMER et GRABOWSKI 1829

Orobanche reticulata subsp. *pallidiflora* (Wimmer et Grabowski) Hayek, *O. cirsii* Fries

Bleiche Distel- (Netzige) Sommerwurz

- **Artbeschreibung**

Meistens kräftige Pflanzen, 10 bis zu 70 (meistens größer als 30) cm hoch. Der Stengel ist kräftig, selten schlank, aufrecht, hell- oder dunkelgelb bis gelblichbraun oder bräunlich gefärbt, spärlich beschuppt (unten dicht, oben lockerer) und zur Gänze reichlich mit Drüsenhaaren besetzt. Die Schuppen sind im unteren Teil dreieckig bis eiförmig und kahl, im oberen Teil lanzettlich und drüsenhaarig, aufrecht bis abstehend. Der Blütenstand ist bei beginnender Blüte dicht- und reichblütig, später vor allem im unteren Teil lockerblütiger und gestreckter; die Blüten sind meistens über den größten Teil des Stengels verteilt. Vorblätter sind nicht vorhanden. Das Tragblatt ist etwa so lang wie die Blütenkrone, rötlich rotbraun, spärlich drüsenhaarig, schmal und abwärts gebogen. Die Kelchhälften sind meist ungespalten, selten ungleich zweizähnig, sehr schwach genervt, eiförmig, etwa halb so lang wie die Blütenkrone und rotbraun bis violett gefärbt. Die Blüten sind mittelgroß, erst aufrecht-abstehend, später fast waagrecht. Die Blütenkrone ist 14 bis 25 mm lang, röhrig oder glockig und über der Ansatzstelle der Staubblätter schwach erweitert, mit spärlichen dunkelvioletten Drüsenhaaren besetzt, an der Basis und in der Mitte weißlich, gelblichweiß bis hellgelb, im Bereich der Oberlippe gelblichbraun, rötlichbraun oder violettpurpurn (die Nerven vorwiegend violett) gefärbt. Die Rückenlinie der Blütenkrone ist vom Grund an gleichmäßig gebogen, oder auf etwa ein Drittel leicht geknickt, in der Mitte fast gerade bis leicht konkav und an der Oberlippe wieder stärker nach vorne gebogen. Die Oberlippe der Blütenkrone ist helmförmig, winklig abfallend und besteht aus zwei sehr breiten, gerundeten, aufrecht abstehenden Lappen. Die Unterlippe der Blütenkrone ist herabgeschlagen, wobei der Mittelzipfel der Unterlippe nur wenig länger ist als die Seitenzipfel, mit gerundeten, meist scharf längsfaltigen, gezähnelten, violett geaderten Lappen. Die Staubblätter sind 2 bis 4 mm hoch über dem Grund der Kronröhre eingefügt und haben dort eine kleine Nektardrüse. Die Staubfäden sind an der Basis kahl oder spärlich behaart, in der Mitte fast kahl und unter den Staubbeuteln mit wenigen Drüsenhaaren besetzt oder kahl. Die Staubbeutel sind an der Naht spärlich behaart. Der Griffel ist spärlich mit Drüsenhaaren besetzt oder fast kahl. Die Narbe besteht aus zwei abgerundeten, kugeligen Lappen und ist braun, braunrot, rotviolett, bräunlichviolett oder purpurn (im oberen Teil meist heller) gefärbt. $2n = 38$.

- **Blütezeit**

Juni bis August, manchmal noch im November. Die Blütezeit dieser Art ist sehr variabel. An verschiedenen Standorten wurden blühenden Pflanzen schon im Mai gefunden, aber ein Jahr später an gleicher Stelle erst im Oktober oder November.

- **Standort**

Orobanche pallidiflora wächst vor allem in wärmeliebenden (Schutt-) Unkrautgesellschaften, in ruderal beeinflußten Halbtrockenrasen und Ackerrändern, an mäßig frischen Ruderalstellen, in Flußwiesen und in Distelgesellschaften auf basen- (kalkhaltigen) und nährstoffreichen, steinig-lehmigen Böden.

- **Wirt**

Schmarotzt vorwiegend auf *Cirsium*- (*C. arvense*, *C. eriophorum*, *C. vulgare* und *C. oleraceum*) und *Carduus*- (*C. acanthoides* und *C. crispus*) Arten

- **Gesamtverbreitung**

Zerstreut bis selten in den niedrigen Lagen Europas (von der planaren bis zur submontanen Stufe), durch Mittel- (sehr selten in den Alpen), Nord- und Osteuropa über die Kaukasusländer bis zum Himalaja. Nördlich bis England (Yorkshire), Dänemark, Südschweden und die baltische Staaten; westlich bis Südwestfrankreich. Die Angaben von Südspanien (Pujadas-Salva *et al.*, 1994) und Nordafrika (Meusel *et al.*, 1978; Hulten *et al.*, 1986) sollten überprüft werden.
Orobanche pallidiflora ist ziemlich verbreitet, aber selten. Sie wächst meistens in größeren Gruppen zusammen.

- **Bemerkungen**

Orobanche pallidiflora ist meistens eine sehr kräftige Pflanze, die vor allem in Unkrautgesellschaften (Ruderalvegetationen) vorkommt. Durch ihre Blütenfarbe (gelblich und violettpurpurne Oberlippe) und die Ausbildung der Kelchhälften, die fast immer ungeteilt sind, ist sie nicht mit anderen Arten zu verwechseln.

Pale Thistle Broomrape

- **Species description**

The plant is usually stout, 10-70 cm tall (usually larger than 30 cm). The stem is stout, rarely slender, erect, light or dark yellow to yellowish-brown or brownish, sparsely scaled (dense below, more laxly above) and richly glandular-pubescent all over. The lower scale leaves are triangular to oval and glabrous, the higher ones are lanceolate, glandular-pubescent and erect to spreading. The inflorescence has numerous flowers in a dense, cylindrical spike in the beginning, later it is lax and elongated, especially in the lower part, with flowers covering the larger part of the stem. Bracteoles are absent. The bract is about as long as the corolla, reddish red-brown, sparsely glandular-pubescent, narrow and deflexed. The calyx-segments are usually entire, rarely unequally bidentate, very faintly veined, oval, about half as long as the corolla and red-brown to violet. The flowers are of medium size, erecto-patent at first, almost horizontal later. The corolla is approximately 14-25 mm long, tubular or campanulate, slightly inflated above the insertion of the stamens, with sparse, dark violet, glandular hairs, whitish at the base and yellowish-white to light yellow in the middle, yellowish-brown, reddish-brown or violet-purple near the upper lip (veins are mainly violet). The dorsal line of the corolla is evenly curved over its entire length or slightly inflected at a third of its length, almost straight to slightly concave in the middle and bent forwards near the upper lip. The upper lip of the corolla is helmet-shaped, flexed at a right angle, and consists of two very broad, rounded, erect to spreading lobes. The lower lip of the corolla is deflexed, with the middle lobe only a little longer than the side lobes, with oval, usually sharply plicate, crenate lobes, which have violet veins. The stamens are inserted 2-4 mm above the base of the corolla-tube, showing a small nectar-gland at that point. The filaments are glabrous or sparsely pubescent at the base, almost glabrous in the middle and sparsely glandular-pubescent or glabrous just below the anthers. The anthers are sparsely pubescent at the line of fusion. The style is sparsely glandular-pubescent or almost glabrous. The stigma consists of two rounded, spherical lobes, which are brown, red-violet, brownish-violet or purple (usually lighter in the upper part). $2n = 38$.

- **Flowering time**

June to August, sometimes as late as November. The flowering time of this species is very variable. In various locations flowering plants have been found in May, while a year later, in the same spot, the plants did not flower until October or November.

- **Habitat**

Orobanche pallidiflora grows preferably in thermophilous herbaceous vegetation on stony ground, in ruderal semi-arid grassland and on the edges of fields, in various ruderal places, in watermeadows and in thistle vegetations on alkaline (calcareous), nutrient-rich, stony-loamy soil.

- **Host**

Parasitic on *Cirsium* species (*C. arvense*, *C. eriophorum*, *C. vulgare* and *C. oleraceum*) and *Carduus* species (*C. acanthoides* and *C. crispus*).

- **Distribution**

Orobanche pallidiflora is rare to sporadic in the low altitude regions of Europe (from the planes to sub-alpine areas), through central Europe (very rare in the Alps), northern and eastern Europe, the Caucasus to the Himalayas. Northward to England (Yorkshire), Denmark, southern Sweden and the Baltic states; westward to south-western France. Findings in southern Spain (Pujadas-Salva *et al.*, 1994) and northern Africa (Meusel *et al.*, 1978; Hulten *et al.*, 1986) should be verified. *Orobanche pallidiflora* is quite widely distributed but rare. It usually grows in fairly large groups.

- **Comments**

Orobanche pallidiflora is usually a very stout plant, growing mainly in herbaceous (ruderal) vegetations. Because of the colour of its flowers (yellowish with a violet-purple upper lip) and the shape of the calyx-segments, which are almost always entire, it cannot be mistaken for any other species.

OROBANCHE PALLIDIFLORA

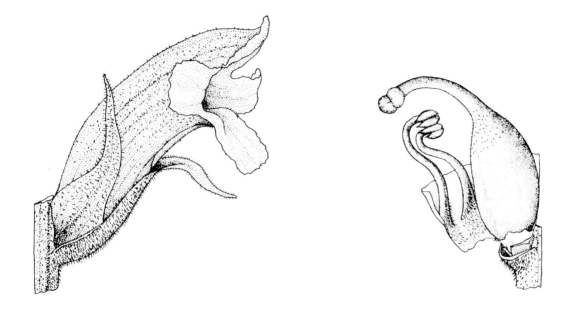

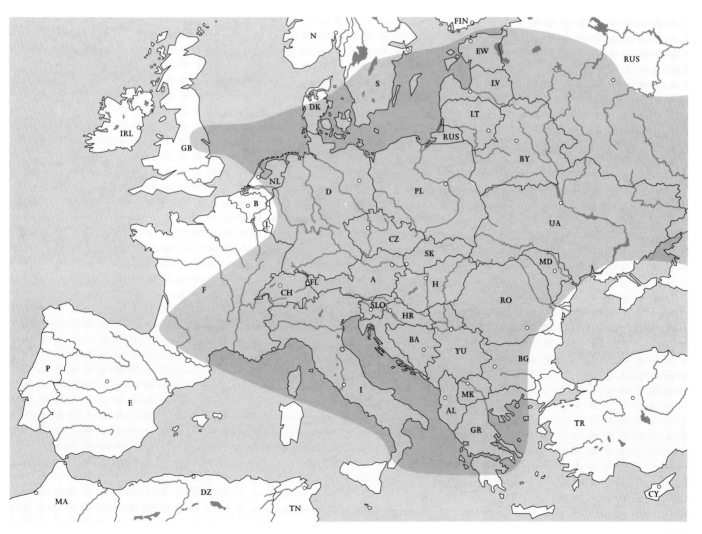

Leuth, Ooijpolder (NL), 26-6-1991

mit Wirtspflanze / with host (*Cirsium oleraceum*), Brilon, Hochsauerland (D), 7-7-1990

Brilon, Hochsauerland (D), 7-7-1990

mit Wirtspflanze / with host (*Cirsium arvense*), Leuth, Ooijpolder (NL), 26-6-1991

Empe, Gelderland (NL), 25-7-1988

4.25

OROBANCHE PICRIDIS — F.W. SCHULTZ ex KOCH 1833
Orobanche carotae Desm.

Bitterkraut-Sommerwurz

- **ARTBESCHREIBUNG**

Pflanzen schlank bis kräftig, etwa 10 bis 70 cm hoch. Der Stengel ist schlank, manchmal auch kräftig, aufrecht, gelb, gelblichweiß (bleich) bis rosa, violett oder rötlichbraun gefärbt, reichlich drüsenhaarig, im unteren Teil locker und bis zum Blütenstand sehr spärlich beschuppt. Die unteren Schuppen sind breitlanzettlich und fast kahl, die oberen schmallanzettlich und mit Drüsenhaaren besetzt, aufrecht bis abstehend. Der Blütenstand ist meist dicht- und reichblütig, zylindrisch, selten arm- und lockerblütig. Vorblätter sind nicht vorhanden. Das Tragblatt ist etwa gleich lang oder länger als die Blütenkrone (die Blütenkrone oft überragend), lanzettlich, an der Basis hellgelb gefärbt mit schwarzbrauner Spitze und ist ab der Mitte herabgeschlagen oder zurückgebogen. Die Kelchhälften sind bis zur Mitte in zwei ungleiche Zähne gespalten, selten aus ungeteilten Hälften bestehend, an der Basis eiförmig und im übrigen Teil schmallanzettlich bis fadenförmig, zwei Drittel bis etwa gleich lang wie die Blütenkrone, an der Basis von gleicher Farbe und an der Spitze meist etwas dunkler (violetter) als die Blütenkrone gefärbt, drüsenhaarig und genervt. Die Blüten sind mittelgroß, anfangs aufrecht-abstehend, später vorwärts gebogen. Die Blütenkrone ist meistens 15 bis 20 mm lang, röhrig, über der Ansatzstelle der Staubblätter schwach bauchig erweitert mit vielen, hellen Drüsenhaaren, die weiß, gelblichweiß oder gelblich gefärbt sind, weißgelblich bis zartrosa (im Bereich der Oberlippe hellrosa oder violett) gefärbt mit rosa oder violetten Adern. Die Rückenlinie der Blütenkrone ist von der Basis an gleichmäßig gebogen, in der Mitte fast gerade oder wenig gebogen und im Bereich der Oberlippe nach vorne gekrümmt, oft mit einem aufgerichteten Spitzchen. Die Oberlippe der Blütenkrone ist ungeteilt, flach ausgerandet oder zweilappig, meistens mit vorgestreckten, abgebogenen oder schwach aufgerichteten Lappen, ihre Zipfel sind reichlich und helldrüsig behaart. Die Unterlippe der Blütenkrone ist schwach herabgeschlagen mit drei fast gleichen, ungleich gezähnelten und längsfaltigen Lappen. Die Staubblätter sind 3 bis 5 mm hoch über dem Grund der Kronröhre eingefügt. Die Staubfäden sind an der Basis schwach oder stark mit Drüsenhaaren besetzt, bis zur Mitte dicht und oben bis zu den Staubbeuteln meist spärlich drüsenhaarig oder kahl. Die Staubbeutel sind an der Naht behaart. Der Griffel ist spärlich mit Drüsenhaaren besetzt. Die Narbe besteht aus zwei zusammenneigenden, halbkugeligen dunkelroten, rosa-violetten, purpurnen oder hellrosa gefärbten Lappen. $2n = 38$.

- **BLÜTEZEIT**

Juni bis Ende Juli, in Südeuropa bereits ab April.

- **STANDORT**

In halbruderalen Rasengesellschaften, ruderal beeinflußten Halbtrockenrasen, in Wiesen, an Straßenrändern und Böschungen auf nährstoffreichen Lehmböden, im Küstenbereich oft in Sanddünen. Die Art wächst vor allem an sonnigen Standorten in den niedrigen Lagen Europas.

- **WIRT**

Schmarotzt auf *Asteraceae*- (am häufigsten auf *Picris hieracioides* und *Crepis*-Arten), selten auf *Daucus carota*.

- **GESAMTVERBREITUNG**

Die Art wächst hauptsächlich in den mediterranen Gebieten, von Portugal und Spanien über den gesamten Mittelmeerraum nach Kleinasien (Westtürkei) bis Transkaukasien. Weiter in großen Teilen von Mittel- und Osteuropa (nördlich bis England, die Niederlande, Dänemark, Südschweden, ehemaliges DDR-Küstengebiet, der Tschechischen Republik und Südpolen).
Die Art ist sehr selten und fehlt in großen Gebieten.

- **BEMERKUNGEN**

Orobanche picridis ist selten. Meistens ist sie durch ihre weißgelbliche bis zartrosa gefärbte Blütenkrone von den anderen Arten zu unterscheiden.

Picris Broomrape

- **SPECIES DESCRIPTION**

The plant is slender to stout, approximately 10-70 cm tall. The stem is slender, sometimes stout, erect, yellow, yellowish-white (pale) to pink, violet or reddish-brown, light or dark yellow to yellowish-brown, richly glandular-pubescent, laxly scaled below, very sparsely scaled above, up to the inflorescence. The lower scale leaves are broadly lanceolate and almost glabrous, the higher ones narrowly lanceolate, glandular-pubescent and erect to spreading. The inflorescence has numerous flowers in a dense, cylindrical spike, rarely lax and with few flowers. Bracteoles are absent. The bract is about as long as or longer than the corolla (often extending beyond the corolla), lanceolate, light yellow at the base, with a brown to black tip, deflexed or recurved from the middle. The calyx-segments are unequally bidentate, deeply bifid, down to the middle, rarely consisting of entire halves, oval at the base and narrowly lanceolate to filiform above, about two thirds of the length of the corolla or as long as the latter. They have the same colour as the corolla below, but are usually slightly darker (more violet) near the tip, and are glandular-pubescent and veined. The flowers are of medium size, erecto-patent at first, flexed forwards later. The corolla is usually 15-20 mm long, tubular, slightly inflated above the insertion of the stamens, with numerous, light, glandular hairs, which are white, yellowish-white or yellowish; the corolla is white-yellowish to pale pink (light pink or violet near the upper lip) and has pink or violet veins. The dorsal line of the corolla is evenly curved over its entire length, almost straight or slightly curved in the middle and flexed forward near the upper lip, often with a raised tip. The upper lip of the corolla is entire, flatly emarginate or bilobate, usually with porrect, flexed or slightly raised lobes, the tips of which are richly glandular-pubescent with bright hairs. The lower lip of the corolla is slightly deflexed, with almost equal, unevenly crenate, plicate lobes. The stamens are inserted 3-5 mm above the base of the corolla-tube. The filaments are sparsely or richly glandular-pubescent at the base, densely glandular-pubescent up to the middle and sparsely glandular-pubescent or glabrous above, up to the anthers. The anthers are pubescent at the line of fusion. The style is sparsely glandular-pubescent. The stigma consists of two hemispherical, dark red, purple or light pink lobes, in close proximity. $2n = 38$.

- **FLOWERING TIME**

June to end of July, as early as April in southern Europe.

- **HABITAT**

Orobanche picridis grows in semi-ruderal grassland, ruderal semi-arid grassland, in meadows, on road-sides and on slopes on nutrient-rich, loamy soil, near the coast often in sand dunes. This species grows mainly in sunny places in the low altitude areas of Europe.

- **HOST**

Parasitic on *Asteraceae* (most frequently on *Picris hieracioides* and *Crepis* species), rarely on *Daucus carota*.

- **DISTRIBUTION**

Orobanche picridis mainly grows in the Mediterranean regions, from Portugal and Spain, through the entire Mediterranean area, to Asia Minor (western part of Turkey) and the Transcaucasian region. It is also found in large parts of central and eastern Europe (northward to England, the Netherlands, Denmark, southern part of Sweden, the coastal area of the former DDR, the Czech Republic and southern Poland).
It is very rare and absent from large areas.

- **COMMENTS**

Orobanche picridis is rare. It is easily distinguished from other species by its white-yellowish to pale pink corolla.

OROBANCHE PICRIDIS

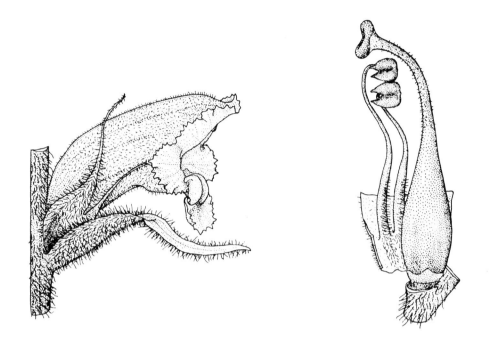

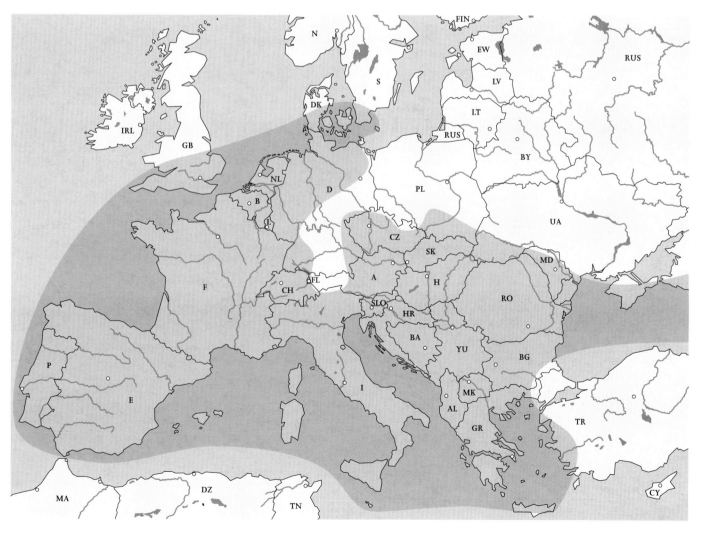

OROBANCHE PICRIDIS

mit Wirtspflanze / with host (*Picris hieracioides*), Wijk aan Zee, Noord-Kennemerland (NL), 20-6-1989

mit Wirtspflanze / with host (*Picris hieracioides*), Wijk aan Zee, Noord-Kennemerland (NL), 20-6-1989

Wijk aan Zee, Noord-Kennemerland (NL), 20-6-1989

Dover, Kent (GB), 15-7-1991

Dover, Kent (GB), 15-7-1991

OROBANCHE RAPUM-GENISTAE　　　　　　　　　　　　　　　　　　THUILLIER 1799
Orobanche major Smith

Greater Broomrape

• SPECIES DESCRIPTION

The plant is usually stout, approximately 30-85 cm tall. The stem is stout, conspicuously swollen below, erect, yellow, yellowish-brown, orange-yellow, brownish or reddish-brown, richly glandular-pubescent, densely scaled (imbricate) below, usually richly scaled above as well. The lower scale leaves are triangular to broadly lanceolate, the higher ones lanceolate, glandular-pubescent and erect to spreading. The inflorescence has numerous flowers in a dense, cylindrical spike, dense above, more lax and elongated in the middle and lower parts, with flowers down to the lower parts of the stem. Bracteoles are absent. The bract is lanceolate, acute, much longer than the corolla, deflexed in the upper part, dark brown (yellowish at the base) and sparsely glandular-pubescent. The calyx consists of two unfused, unequally bidentate, rarely entire halves with elongated acute tips, richly glandular-pubescent, shorter than or half as long as the corolla and of the same colour (or slightly lighter at the base) as the latter. The flowers are large, erect to spreading. The corolla is 20-25 mm long, widely tubular to campanulate, inflated above the insertion of the stamens, with numerous glandular hairs, light yellowish, reddish-brown to violet-brown or dark brown, with dark red veins. The dorsal line of the corolla is evenly curved over its entire length and helmet-shaped near the top. The upper lip of the corolla is helmet-shaped, entire, with porrect or slightly raised lobes. The lower lip of the corolla is deflexed, sparsely glandular-pubescent (almost glabrous) with almost equal (middle lobe somewhat larger), rounded, crenate lobes. The stamens are inserted up to 2 mm above the base of the corolla-tube. The filaments are glabrous at the base, densely glandular-pubescent just below the anthers. The anthers are almost glabrous. The style is usually glandular-pubescent above. The stigma consists of two spherical lobes and is wax-coloured or (golden) yellow. The flowers have an unpleasant smell. $2n = 38$.

• FLOWERING TIME

Mid-May to September, depending on altitude.

• HABITAT

Many locations are in sunny, warm heathland with furze, in open oak and birch forests, in oak forests with acidified soil and in nutrient-poor pastures on loamy and sandy soil poor in calcium. *Orobanche rapum-genistae* frequently grows in large, close groups.

• HOST

Parasitic on *Fabaceae* species (in particular on *Cytisus scoparius*, more rarely on *Genista tinctoria* or on *Ulex* species).

• DISTRIBUTION

Western Europe; northward to southern Scotland, the Netherlands and northern Germany; eastward to western Germany, the western part of Switzerland and Italy; southward to Portugal, Spain and Sicily. On Corsica and Sardinia. Also in north-western Africa.
Orobanche rapum-genistae is rare, but has a rather wide range and grows in clusters in some areas.

• COMMENTS

The variety '*hypoxantha*' Beck 1890 has been described. Stem and corolla of this form are purely yellow.
Orobanche rapum-genistae is very stout and can be recognized mainly by its lax and very long inflorescence. The bract is usually much longer than the corolla and is deflexed in the upper part.

OROBANCHE RAPUM-GENISTAE

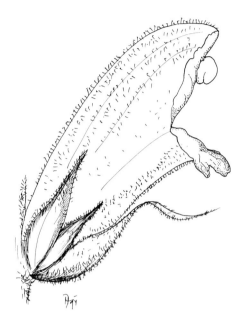
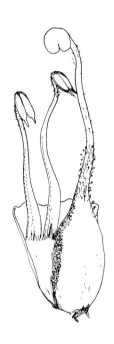

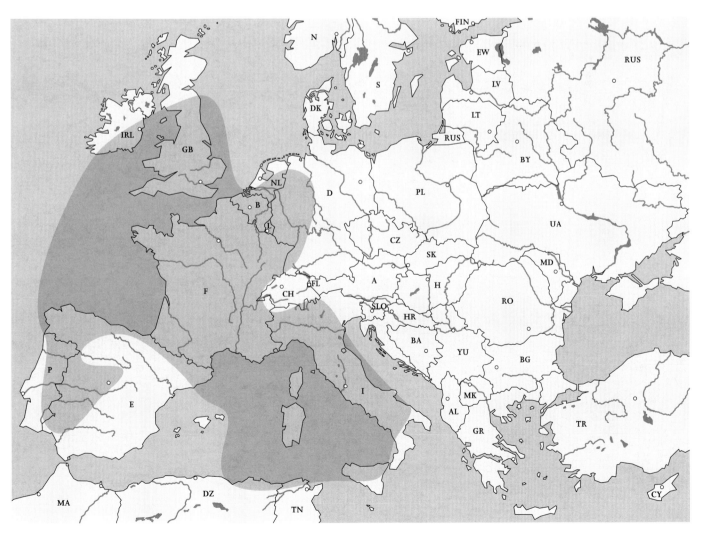

OROBANCHE RAPUM-GENISTAE

mit Wirtspflanze / with host (*Cytisus scoparius*), Kerschenbach, Eifel (D), 7-7-1983

mit Wirtspflanze / with host (*Cytisus scoparius*), Kerschenbach, Eifel (D), 30-6-1989

Chooz, Ardennes (F), 3-6-1989

Montfort, Limburg (NL), 4-6-1989

Kerschenbach, Eifel (D), 7-7-1983

OROBANCHE RETICULATA ──────────────────── WALLROTH 1825
Orobanche platystigma Reichenbach; *O. scabiosae* W.D. Koch; *O. cardui* Sauter

Eigentliche Distel- (Netzige) Sommerwurz

- **ARTBESCHREIBUNG**

Pflanzen meistens klein, 10 bis 20 cm hoch. Es kommen aber auch kräftige Pflanzen vor, die bis zu 70 cm hoch werden. Der Stengel ist schlank, manchmal kräftig, aufrecht, gelblich- bis violettbraun oder bräunlich gefärbt, spärlich beschuppt (unten dicht, oben lockerer), zur Gänze reichlich mit Drüsenhaaren besetzt. Die Schuppen sind im unteren Teil dreieckig bis lanzettlich und kahl, im oberen Teil lanzettlich und drüsenhaarig, aufrecht bis abstehend. Der Blütenstand ist meistens zylindrisch, dicht- und armblütig, später bei größeren Exemplare vor allem im unteren Teil lockerblütiger und gestreckter; die Blüten sind meistens über den oberen Teil der Stengel verteilt. Vorblätter sind nicht vorhanden. Das Tragblatt ist meistens länger als die Blütenkrone, lanzettlich, dunkelbraun gefärbt, drüsenhaarig, und manchmal ab der Mitte abwärts gebogen. Die Kelchhälften sind meist ungespalten, selten ungleich zweizähnig, sehr schwach genervt, lanzettlich, im unteren Teil eiförmig, etwa halb so lang wie die Blütenkrone und braun, schwarzbraun, rotbraun oder violett gefärbt. Die Blüten sind mittelgroß, erst aufrecht-abstehend, später fast waagrecht. Die Blütenkrone ist 14 bis 25 mm lang, röhrig oder glockig und über der Ansatzstelle der Staubblätter schwach erweitert, reichlich (dicht) mit dunkelvioletten Drüsenhaaren besetzt, nur an der Basis gelblich, und in der Mitte und im Bereich der Oberlippe überwiegend mehr oder weniger intensiv violett oder purpurviolett (die Nerven dunkelviolett) gefärbt. Die Rückenlinie der Blütenkrone ist vom Grund an gleichmäßig gebogen, oder auf etwa ein Drittel leicht geknickt, in der Mitte fast gerade bis leicht konkav und an der Oberlippe wieder stärker nach vorne gebogen. Die Oberlippe der Blütenkrone ist helmförmig, winklig abfallend und besteht aus zwei sehr breiten, gerundeten, aufrecht abstehenden Lappen. Die Unterlippe der Blütenkrone ist herabgeschlagen, wobei der Mittelzipfel nur wenig länger oder gleich lang wie die Seitenzipfel ist, mit gerundeten, meist scharf längsfaltigen, gezähnelten, violett geaderten Lappen. Die Staubblätter sind 2 bis 4 mm hoch über dem Grund der Kronröhre eingefügt und haben dort eine kleine Nektardrüse. Die Staubfäden sind an der Basis kahl oder spärlich behaart, in der Mitte fast kahl und unter den Staubbeuteln mit Drüsenhaaren besetzt. Die Staubbeutel sind an der Naht spärlich behaart. Der Griffel ist spärlich mit Drüsenhaaren besetzt oder fast kahl. Die Narbe besteht aus zwei abgerundeten, kugeligen Lappen und ist braun, braunrot, rotviolett, bräunlichviolett oder purpurn (im oberen Teil meist heller) gefärbt. $2n = 38$.

- **BLÜTEZEIT**

Juni bis September.

- **STANDORT**

Orobanche reticulata wächst hauptsächlich im Gebirgssteinrasen, in wärmeliebenden (Schutt-) Unkrautgesellschaften, in subalpinen Blaugrasrasen, auf mäßig frischen Ruderalstellen, in Rasengesellschaften, in Distelgesellschaften auf basen- (kalkhaltigen) und nährstoffreichen, steinig-lehmigen Böden.

- **WIRT**

Schmarotzt vorwiegend auf *Cirsium erisithales, Carlina acaulis, Scabiosa lucida, Carduus defloratus* und *Knautia dipsacifolia*.

- **GESAMTVERBREITUNG**

Verbreitet bis zerstreut in den höheren Lagen der Alpen und des Alpenvorlandes von Nordostspanien, Frankreich (auch in den Vogesen) durch Mittel- und Osteuropa über die Kaukasusländer bis zum Himalaja. Südwärts bis zu den Pyrenäen (Spanien), Südfrankreich, Italien (Abruzzen), den Balkanländern und Griechenland (Peloponnes).
Die Art ist ziemlich verbreitet, aber selten.

- **BEMERKUNGEN**

Von *Orobanche reticulata* wurde die Varietät '*procera*' (Koch) Beck 1890 beschrieben. Im Gegensatz zu *O. reticulata* ist deren Blumenkrone weit röhrig-glockig, 15 bis 18 mm lang, weiß bis gelblich gefärbt mit violetten Drüsenhaaren und die Rückenlinie zur Gänze gleichmäßig gekrümmt. Diese Varietät schmarotzt vor allem auf *Carduus*- und *Cirsium*-Arten. Ihr Verbreitungsgebiet liegt vor allem im Oberrheingebiet, dem Osten Deutschlands und in den osteuropäischen Staaten.
Orobanche reticulata ist vor allem in den zentral- und südeuropäischen Gebirgen (Alpen, Karpaten, Apeninnen, Balkan-Halbinsel) und im Kaukasus verbreitet (montan bis subalpin). Durch ihre Blütenfarbe (nur an der Basis gelb und im übrigen Teil violettpurpurn gefärbt), ihre dicht drüsenhaarige Blumenkrone und die Ausbildung der Kelchhälften, die fast immer ungeteilt sind, ist sie nicht mit anderen Arten zu verwechseln.

Thistle Broomrape

- **SPECIES DESCRIPTION**

The plant is usually small, 10-20 cm tall. Sometimes stout plants of up to 70 cm are found. The stem is slender, sometimes robust, erect, yellowish-brown to violet-brown or brownish, sparsely scaled, (densely scaled below, more sparsely scaled above), richly glandular-pubescent all over. The lower scale leaves are triangular to lanceolate and glabrous, the higher ones are lanceolate, glandular-pubescent and erect to spreading. The inflorescence has few flowers in a dense, cylindrical spike, spikes on larger plants are more lax and elongated in the middle and lower parts later on; flowers are usually on the upper part of the stem. Bracteoles are absent. The bract is usually longer than the corolla, lanceolate, dark brown, glandular-pubescent and sometimes flexed downwards from the middle. The calyx-segments are usually entire, rarely unequally bidentate, very lightly veined, lanceolate, oval below, about half as long as the corolla and brown to black, red-brown or violet. The flowers are of medium size, erecto-patent at first and almost horizontal later. The corolla is 14-25 mm long, tubular or campanulate, slightly inflated above the insertion of the stamens, with numerous, dense, dark violet glandular hairs, yellowish only at the base, predominantly more or less intensely violet or purple-violet in the middle and near the upper lip (veins are dark violet). The dorsal line of the corolla is evenly curved over its entire length or slightly flexed at a third of its length, almost straight to slightly concave in the middle and more strongly curved forwards near the upper lip. The upper lip of the corolla is helmet-shaped, flexed at right angles and consisting of two very broad, rounded, erecto-patent lobes. The lower lip of the corolla is deflexed, the middle lobe is a little longer than or as long as the side lobes; with rounded, usually sharply plicate, crenate lobes with violet veins. The stamens are inserted 2-4 mm above the base of the corolla-tube and have a small nectar-gland at the insertion. The filaments are glabrous or sparsely pubescent at the base, almost glabrous in the middle and glandular-pubescent just below the anthers. The anthers are sparsely glandular-pubescent at the line of fusion. The style is sparsely glandular-pubescent or almost glabrous. The stigma consists of two rounded, spherical lobes and is brown, brown-red, red-violet, brownish-violet or purple (mostly lighter in the upper part). $2n = 38$.

- **FLOWERING TIME**

June to September.

- **HABITAT**

Orobanche reticulata grows mainly in stony, alpine grassland, in thermophilous herbaceous vegetation (on rubble), among sedges and rushes in subalpine regions, in ruderal vegetations, in grassy or thistle vegetations, on alkaline (calcareous) and nutrient-rich, stony-loamy soil.

- **HOST**

Parasitic mainly on *Cirsium erisithales, Carlina acaulis, Scabiosa lucida, Carduus defloratus* and *Knautia dipsacifolia*.

- **DISTRIBUTION**

Frequent to sporadic in the higher altitude regions of the Alps and the alpine foothills of north eastern Spain, France (also in the Vosges), through central and eastern Europe and the Caucasus to the Himalayas. Southward to the Pyrenees (Spain), southern France, Italy (Abruzzi), the Balkan countries and Greece (Peloponnesus).
This species has a fairly wide range, but is rare.

- **COMMENTS**

A variety '*procera*' (Koch) Beck 1890 has been identified. In contrast to *Orobanche reticulata* the variety '*procera*' has a widely tubular-campanulate corolla 15-18 mm long, white to yellowish with violet glandular hairs and with an evenly curved dorsal line. It is parasitic mainly on *Carduus* and *Cirsium* species. Its range is mainly in the upper Rhine area, eastern Germany and the eastern European states.
Orobanche reticulata grows mainly in the mountains of central and southern Europe (the Alps, Carpathian Mountains, Apennine Mountains, the Balkans) and in the Caucasus (montane to sub-alpine). *Orobanche reticulata* cannot be mistaken for any other species, because of the colour of its flowers (yellow only at the base, violet-purple in other parts), its densely glandular-pubescent corolla and its calyx-segments, which are nearly always entire.

OROBANCHE RETICULATA

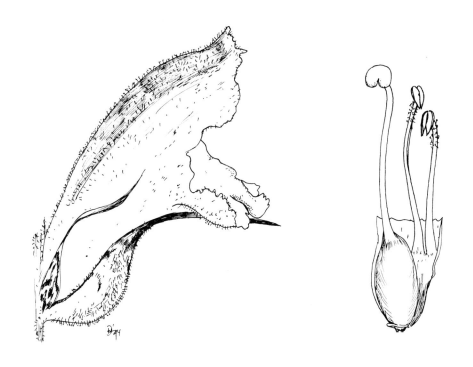

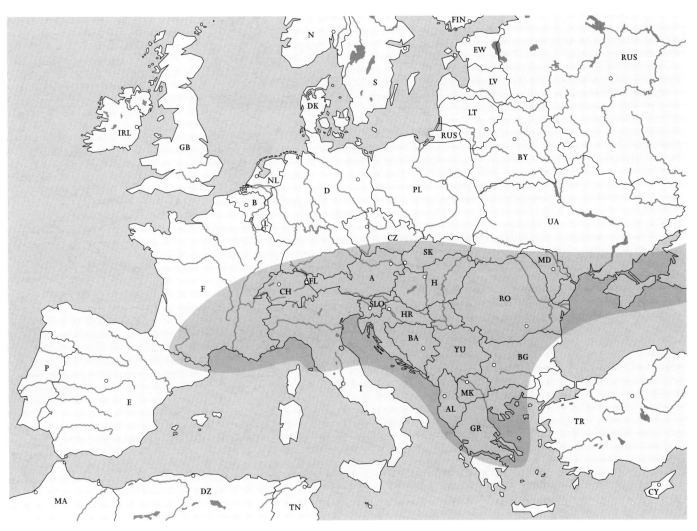

OROBANCHE RETICULATA

mit Wirtspflanze / with host (*Carduus defloratus*), Neuberg an der Mürz, Alpen Rax Schneeberg-Gebiet (A), 9-7-1994

mit Wirtspflanze / with host (*Carduus defloratus*), Neuberg an der Mürz, Alpen Rax Schneeberg-Gebiet (A), 9-7-1994

Neuberg an der Mürz, Alpen Rax Schneeberg-Gebiet (A), 9-7-1994

Neuberg an der Mürz, Alpen Rax Schneeberg-Gebiet (A), 9-7-1994

Neuberg an der Mürz, Alpen Rax Schneeberg-Gebiet (A), 9-7-1994

4.28

OROBANCHE SALVIAE — F.W. SCHULTZ ex KOCH 1833

Salbei-Sommerwurz

- **Artbeschreibung**

Pflanzen meist kräftig oder schlank, etwa 10 bis 55 cm hoch. Der Stengel ist schlank, manchmal auch kräftig, aufrecht, gelb, gelbbraun oder braun gefärbt, spärlich bis reichlich mit hellen Drüsenhaaren besetzt, bis zum Blütenstand unten reichlich und dicht (dachig), oben spärlicher und lockerer beschuppt. Die unteren Schuppen sind dreieckig bis lanzettlich und fast kahl, die oberen breit- oder schmallanzettlich, mit Drüsenhaaren besetzt, aufrecht bis abstehend. Der Blütenstand ist vorwiegend dichtblütig und zylindrisch, später im unteren Teil lockerblütiger, selten zur Gänze lockerblütig; der Blütenstand ist etwa ein Drittel so lang wie der Stengel. Vorblätter sind nicht vorhanden. Das Tragblatt ist etwas länger als die Blütenkrone, gelbbraun gefärbt, lanzettlich mit schwarzbrauner, im oberen Teil herabgeschlagener Spitze, drüsenhaarig. Die Kelchhälften sind ungespalten oder ungleich zweizähnig, schwach genervt, meist nur halb so lang und oft heller (rötlicher) als die Blütenkrone gefärbt. Die Blüten sind mittelgroß, anfangs aufrecht-abstehend, später vorwärts gebogen bis fast waagrecht-abstehend. Die Blütenkrone ist meistens 15 bis 23 mm lang, röhrig, über der Ansatzstelle der Staubblätter schwach bauchig erweitert mit vielen, hellen Drüsenhaaren, gelbbraun, braun, manchmal auch rötlichbraun gefärbt. Die Rückenlinie der Blütenkrone ist von der Basis an meistens gleichmäßig gebogen, manchmal kurz vor der Oberlippe fast gerade oder stärker nach vorne gekrümmt. Die Oberlippe der Blütenkrone ist ungeteilt, flach ausgerandet oder zweilappig, meistens mit vorgestreckten oder schwach aufgerichteten Lappen, ihre Zipfel sind drüsig behaart. Die Unterlippe der Blütenkrone ist wenig herabgeschlagen, oft nur vorgestreckt, mit drei gezähnelten, abgerundeten Lappen, wobei der Mittelzipfel nur wenig länger oder gleich lang wie die beiden Seitenzipfel ist. Die Staubblätter sind 3 bis 5 mm hoch über dem Grund der Kronröhre eingefügt. Die Staubfäden sind am Grund bis zur Mitte reichlich behaart und oben bis zu den Staubbeuteln meist spärlich drüsenhaarig oder fast kahl. Die Staubbeutel sind an der Naht behaart. Der Griffel ist reichlich mit Drüsenhaaren besetzt. Die Narbe besteht aus zwei zusammenneigenden, goldgelben oder orange gefärbten Lappen, später bräunlich. $2n = 38$.

- **Blütezeit**

Juli bis September. Eine spätblühende Art.

- **Standort**

Viele Standorte befinden sich in edellaubholzreichen Schlucht- und in staudenreichen Bergmischwäldern auf basenreichen Lehmböden.

- **Wirt**

Schmarotzt auf *Salvia*-Arten (meist *Salvia glutinosa*).

- **Gesamtverbreitung**

Mitteleuropa, dort in den Alpen und Alpenvorland (Südostfrankreich, die Schweiz, Süddeutschland, Liechtenstein, Österreich, Norditalien, Ungarn, Bosnien-Herzegowina, Kroatien und Slowenien). Die Angaben von Rumänien (Savulescu et al., 1963) sollten überprüft werden.
Orobanche salviae ist selten und sehr anspruchsvoll. In Deutschland, wo sie in den Voralpen und Alpengebieten vorkommt, sind nur noch wenige Fundorte bekannt.

- **Bemerkungen**

Orobanche salviae wächst meistens nur in wenigen Exemplaren auf schattigen Standorten. Weil sie nur auf einem Wirt schmarotzt, der übrigens weit entfernt stehen kann, wegen ihrer späten Blütezeit und ihres gelbbraun gefärbten Blütenstandes ist sie von anderen *Orobanche*-Arten gut zu unterscheiden.

Sage Broomrape

- **Species description**

The plant is usually stout or slender, 10-55 cm tall. The stem is slender, sometimes stout, erect, yellow, yellowish-brown or brown, sparsely to richly glandular-pubescent with light glandular hairs, densely scaled (imbricate) below, sparsely and laxly scaled above, up to the inflorescence. The lower scale leaves are triangular to lanceolate and almost glabrous, the higher ones are broadly or narrowly lanceolate, glandular-pubescent and erect to spreading. The inflorescence usually has a dense, cylindrical spike, more lax in the lower part later on, rarely lax over its entire length; the inflorescence covers about one third of the stem. Bracteoles are absent. The bract is usually slightly longer than the corolla, yellow-brown, lanceolate with brown to black tip, which is glandular-pubescent and deflexed above. The calyx-segments are entire or unequally bidentate, lightly veined, about half as long as the corolla and often more lightly coloured (reddish) than the corolla. The flowers are of medium size, erecto-patent at first and flexed forwards to almost horizontal later. The corolla is usually 15-23 mm long, tubular, slightly inflated above the insertion of the stamens, with numerous, light, glandular hairs, yellow-brown, brown, sometimes reddish-brown. The dorsal line of the corolla is usually evenly curved over its entire length, sometimes almost straight or more flexed forwards near the upper lip. The upper lip of the corolla is entire, flatly emarginate or bilobate, usually with porrect or slightly raised lobes, the tips glandular-pubescent. The lower lip of the corolla is slightly deflexed, often just porrect, with three crenate, rounded lobes; the middle lobe is a little longer than or as long as the side lobes. The stamens are inserted up to 3-5 mm above the base of the corolla-tube. The filaments are richly pubescent up to the middle and sparsely glandular-pubescent or almost glabrous above, up to the anthers. The anthers are pubescent at the line of fusion. The style is richly glandular-pubescent. The stigma consists of two golden yellow or orange lobes, in close proximity. $2n = 38$.

- **Flowering time**

July to September. A late-flowering species.

- **Habitat**

Many locations are in deciduous forests in ravines or in mixed alpine forests with a well-developed herbaceous undergrowth, on alkaline, loamy soil.

- **Host**

Parasitic on *Salvia* species (usually on *Salvia glutinosa*).

- **Distribution**

Central Europe, in the Alps and alpine foothills (south-eastern France, Switzerland, southern Germany, Liechtenstein, Austria, northern Italy, Hungary, Bosnia-Hercegovina, Croatia and Slovenia). Findings in Rumania (Savulescu et al., 1963) should be verified.
Orobanche salviae is rare and very exacting. Only a few locations remain in Germany, where the plant grows in the alpine foothills and the Alps.

- **Comments**

Orobanche salviae usually grows in small numbers in shaded places. It is easily distinguished from other *Orobanche* species because it is parasitic on only one host (which may actually grow at a considerable distance), and through its late flowering time and its yellow-brown inflorescence.

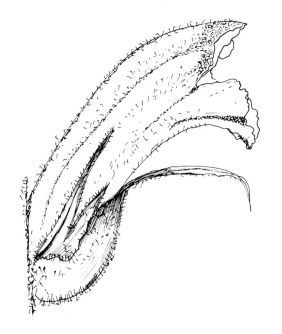
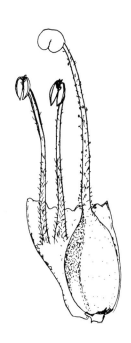
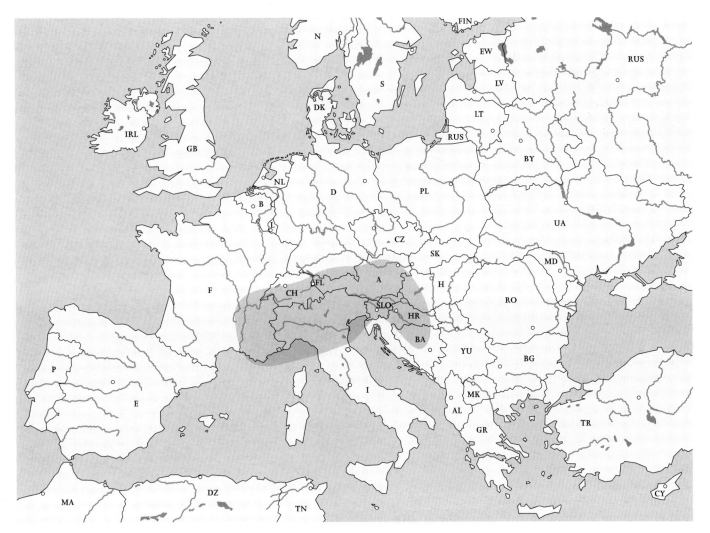

mit Wirtspflanze / with host (*Salvia glutinosa*), Marktschellenberg, Berchtesgadener Alpen (D), 29-7-1993

Marktschellenberg, Berchtesgadener Alpen (D), 29-7-1993

Marktschellenberg, Berchtesgadener Alpen (D), 29-7-1993

Marktschellenberg, Berchtesgadener Alpen (D), 29-7-1993

Marktschellenberg, Berchtesgadener Alpen (D), 29-7-1993

4.29

OROBANCHE TEUCRII — HOLANDRE 1829
Orobanche atrorubens F.W. Schultz

Gamander-Sommerwurz | # Germander Broomrape

- **ARTBESCHREIBUNG**

Die Pflanzen sind meistens klein, relativ kräftig und erreichen eine Größe von etwa 15 bis 40 cm. Der Stengel ist schlank und aufrecht, hellbraun, rötlichbraun oder gelbbraun gefärbt, selten zur Gänze gelb oder bleichgelb, reichlich und dicht mit hellen Drüsenhaaren besetzt, unten dicht und oben bis zur Hälfte der Stengel lockerer mit Schuppen besetzt. Die Schuppen sind im unteren Teil dreieckig bis eiförmig, spärlich drüsenhaarig bis kahl, im oberen Teil lanzettlich und spärlich drüsenhaarig, dunkelbraun, aufrecht bis abstehend (die oberen Schuppen manchmal etwas nach unten gebogen). Der Blütenstand ist meist lockerblütig, selten dichtblütig, kurz mit wenigen ziemlich großen Blüten (bei großen Exemplaren im unteren Teil lockerblütiger). Vorblätter sind nicht vorhanden. Das Tragblatt ist etwa gleich lang oder etwas kürzer als die Blütenkrone, drüsenhaarig, ab der Mitte abwärts gebogen, spitz, gelblichbraun oder braun gefärbt. Die Kelchhälften sind meist bis zur Mitte in zwei fast gleiche Zähne gespalten, etwa ein Drittel bis halb so lang wie die Blütenkrone, spärlich oder reichlich mit hellen Drüsenhaaren besetzt und meist heller als die Blütenkrone gefärbt. Die Blüten sind relativ groß, aufrecht bis abstehend, mit einigen dunkler gefärbten Nerven. Die Blütenkrone ist etwa 20 bis 30 mm lang, über der Ansatzstelle der Staubblätter bauchig erweitert mit hellen, kurzen Drüsenhaaren, außen bräunlichlila bis rötlich (zur Basis hin heller) gefärbt mit dunkelroten (selten violetten) oder dunkelbraunen Nerven, innen meist heller (hell- bis bleichgelb), selten zur Gänze strohgelb gefärbt. Die Rückenlinie der Blütenkrone ist im untersten Teil leicht gebogen, in der Mitte fast gerade und nahe der Oberlippe stark nach vorne gekrümmt. Die Oberlippe der Blütenkrone ist fast ungeteilt oder etwas ausgerandet mit aufgerichteten Lappen und spärlich mit dunkelvioletten oder hellen Drüsenhaaren besetzt. Die Unterlippe der Blütenkrone besteht aus drei fast gleichgroßen gezähnelten, herabgebogenen, violett geaderten Lappen, die kahl (selten spärlich drüsenhaarig) sind. Die Staubblätter sind 3 bis 5 mm hoch über dem Grund der Kronröhre eingefügt, am Grund mit goldgelber Halbmonddrüse umgeben. Die Staubfäden sind unten sehr dicht behaart und oben bis zu den Staubbeuteln reichlich mit Drüsenhaaren besetzt. Die Staubbeutel sind meistens an der Naht behaart. Der Griffel ist reichlich mit Drüsenhaaren besetzt. Die Narbe besteht aus zwei Lappen und ist dunkelbraun, purpurn bis orange oder strohgelb (bei gelben Exemplaren) gefärbt. 2n = 38.

- **BLÜTEZEIT**

Ende Mai bis Juli. Eine relativ frühblühende Art.

- **STANDORT**

Orobanche teucrii wächst vor allem auf kurzrasigen Trocken- und Halbtrockenrasen (steinige, trockene und sonnige Stellen) und Geröllhalden auf kalkreichen Lehm- oder Lößböden.

- **WIRT**

Auf *Teucrium*-Arten (besonders *Teucrium chamaedrys* und *T. montanum*) schmarotzend.

- **GESAMTVERBREITUNG**

Vor allem Mitteleuropa; von den Pyrenäen bis zu den Karpaten. Nördlich bis Belgien, Mitteldeutschland, Österreich und die Slowakei; südlich bis Nordostspanien, Norditalien und den nördlichen Teil der Balkanhalbinsel; östlich bis Rumänien.
Die Art ist selten und tritt zerstreut auf.

- **BEMERKUNGEN**

Selten ist die Form 'aurea' Teyber 1913, die zum Beispiel an einigen Stellen in der Eifel (Deutschland) vorkommt. Ihre Blumenkrone ist gänzlich goldgelb gefärbt.
Überwiegend sind die Pflanzen von *Orobanche teucrii* klein. Sie besitzt wenige und ziemlich große Blüten. Sie wächst hauptsächlich in Trockenrasen an steinigen Stellen. *O. teucrii* ist leicht verwechselbar mit *O. alba*, aber *O. alba* ist mit ihrer auffallenden Farbe der Blumenkrone, ihren deutlich violetten Adern und der Ausbildung der Kelchhälften, die fast immer ungeteilt sind, von dieser zu unterscheiden.

- **SPECIES DESCRIPTION**

The plant is usually small, relatively robust, reaching approximately 15-40 cm. The stem is slender and erect, light brown, reddish-brown or yellow-brown, rarely yellow or pale yellow, richly and densely glandular-pubescent, densely scaled below and laxly scaled above, up to the middle of the stem. The lower scale leaves are triangular to oval, sparsely glandular-pubescent to glabrous, the upper scale leaves are lanceolate and sparsely glandular-pubescent, dark brown, erect to spreading (upper scale leaves are sometimes flexed downwards). The inflorescence is usually short, lax or rarely dense, with few, rather large flowers (lax below on large plants). Bracteoles are absent. The bract is about as long as or slightly shorter than the corolla, glandular-pubescent, sometimes deflexed from the middle, acute and yellowish-brown or brown. The calyx-segments are usually bifid down to the middle, with almost equal halves, about a third or half as long as the corolla, sparsely or richly covered with light glandular hairs, its colour usually lighter than that of the corolla. The flowers are relatively large, erect to spreading, with a few darkly coloured veins. The corolla is approximately 20-30 mm long, inflated above the insertion of the stamens, with light, short glandular hairs, brownish-violet to reddish on the outside (lighter near the base) with dark red (rarely violet) or dark brown veins; usually lighter (bright or pale yellow) on the inside, rarely entirely straw-yellow. The dorsal line of the corolla is slightly curved in the lower part, almost straight in the middle part and strongly flexed forwards near the upper lip. The upper lip of the corolla is almost entire or slightly emarginate, with raised lobes, with sparse dark violet or light glandular hairs. The lower lip of the corolla consists of three almost equal, glabrous (rarely sparsely glandular-pubescent), crenate, deflexed lobes, with violet veins. The stamens are inserted 3-5 mm above the base of the corolla-tube, with golden yellow, crescent-shaped glands at the base. The filaments are very densely pubescent at the base and richly glandular-pubescent above, up to the anthers. The anthers are usually pubescent at the line of fusion. The style is richly glandular-pubescent. The stigma consists of two lobes and is dark brown, purple to orange or straw-yellow (on yellow plants). 2n = 38.

- **FLOWERING TIME**

End of May to July. This species flowers relatively early in the year.

- **HABITAT**

Orobanche teucrii grows mainly in arid and semi-arid grassland (with short turf) in stony, dry and sunny places and on screes, on calcareous loamy soil or loess.

- **HOST**

Parasitic on *Teucrium* species (especially on *Teucrium chamaedrys* and *T. montanum*).

- **DISTRIBUTION**

Mainly in central Europe; from the Pyrenees to the Carpathian mountains. Northward to Belgium, central Germany, Austria and the Slovak Republic; southward to north-eastern Spain, northern Italy and the northern part of the Balkans; eastward to Rumania.
The species is rare and sporadic.

- **COMMENTS**

The variety 'aurea' Teyber 1913 is very rare, growing, for instance, in a few places in the Eifel (Germany). Its corolla is entirely golden yellow.
Orobanche teucrii is predominantly small with few, rather large flowers. It grows mainly in dry grassland in stony places. It is difficult to distinguish *O. teucrii* from *O. alba*, but the distinct colour of the corolla of the latter, its clearly violet veins and the shape of the calyx-segments, which are almost always entire, allow identification.

OROBANCHE TEUCRII

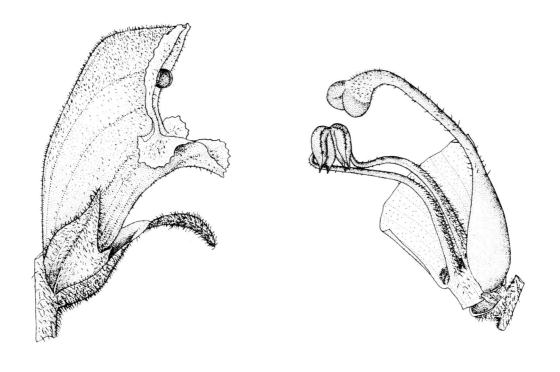

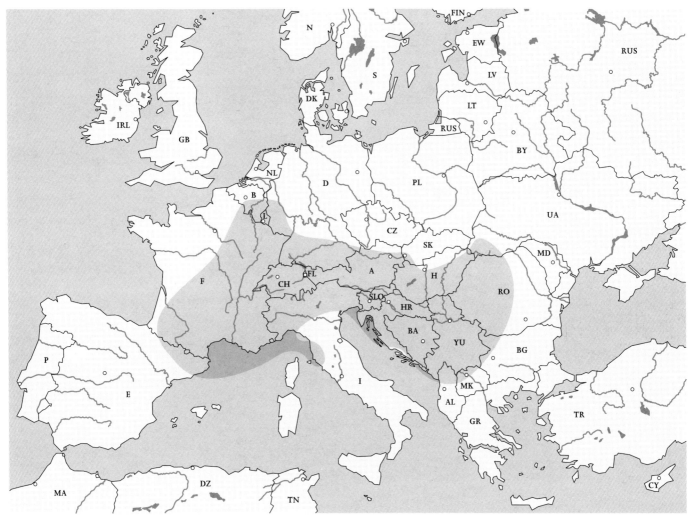

mit Wirtspflanze / with host (*Teucrium chamaedrys*), Gilsdorf, Eifel (D), 18-6-1988 (f. *aurea*)

Gilsdorf, Eifel (D), 18-6-1988

Vogtsburg, Kaiserstuhl (D), 5-5-1990

Gilsdorf, Eifel (D), 18-6-1988 (f. *aurea*)

Ripsdorf, Eifel (D), 14-7-1985

4.30

OROBANCHE VARIEGATA — WALLROTH 1825
Orobanche spartii Vaucher

Bunte Sommerwurz

• ARTBESCHREIBUNG
Die Pflanzen sind meistens kräftig, etwa 30 bis 70 cm hoch. Der Stengel ist kräftig, aufrecht, gelb, gelblichbraun, orangegelb, rötlich, goldgelb, selten purpurn gefärbt, reichlich (selten spärlich) mit Drüsenhaaren besetzt (manchmal fast kahl), unten dicht (dachig), im oberen Teil lockerer beschuppt. Die unteren Schuppen sind breitlanzettlich bis eiförmig und kahl, die oberen lanzettlich und drüsenhaarig, aufrecht bis abstehend. Der Blütenstand ist zylindrisch, dicht- und reichblütig, am Grunde oft lockerblütig; die Blüten sind meistens über zwei Drittel der Stengel verteilt. Vorblätter sind nicht vorhanden. Das Tragblatt ist etwa so lang oder länger als die Blütenkrone, lanzettlich und abwärts gebogen, gelblichbraun gefärbt und fast kahl. Die Kelchhälften sind mehr oder weniger ungleich zweizähnig, selten ungeteilt, etwa halb bis zwei Drittel so lang wie die Blütenkrone, schmallanzettlich, drüsenhaarig (vor allem an der Basis), an der Spitze oft dunkler gefärbt als die Blütenkrone. Die Blüten sind mittelgroß, erst aufrecht, später (waagrecht) abstehend. Die Blütenkrone ist 14 bis 23 mm lang, glockenförmig und über der Ansatzstelle der Staubblätter stark bauchig erweitert, außen mit kurzen Drüsenhaaren besetzt, innen kahl, außen hell- bis dunkelgelb (am Zipfel rötlich oder rotbraun), innen dunkelrot gefärbt, schwach genervt. Die Rückenlinie der Blütenkrone ist vom Grund an gleichmäßig gebogen (sie ist in der Mitte meistens etwas stärker gekrümmt) und im Bereich der Oberlippe fast gerade. Die Oberlippe der Blütenkrone ist meistens ungeteilt oder zweilippig, mit breiten, vorgestreckten, gerundeten Lappen. Die Unterlippe der Blütenkrone ist herabgeschlagen mit drei gerundeten, ungleich-gezähnelten Lappen, wobei der Mittellappen meistens doppelt so groß ist wie die beiden Seitenlappen. Die Staubblätter sind 2 bis 4 mm hoch über dem Grund der Kronröhre eingefügt. Die Staubfäden sind im unteren Teil mehr oder weniger behaart (oder kahl) und oben bis zu den Staubbeuteln reichlich bis spärlich mit Drüsenhaaren besetzt. Die Staubbeutel sind an der Naht kurz behaart. Der Griffel ist im unteren Teil (Fruchtknoten) kahl und im Bereich der Narbe spärlich mit Drüsenhaaren besetzt. Die Narbe besteht aus zwei abgerundeten, kugeligen Lappen und ist im unteren (dunkel)gelb und im oberen Teil rötlich gefärbt. Die Blüten riechen unangenehm. $2n = 38$.

• BLÜTEZEIT
Juni und Juli, im mediterranen Bereich schon ab Mitte April.

• STANDORT
Orobanche variegata wächst hauptsächlich auf Trocken-, Halbtrocken- und Xerothermrasen, auch an lichten Gebüschsäumen (Waldrändern) und an steinig-felsigen Kalkhängen an warmen, sonnigen Standorten auf basenreichen Lehm- und Kalkböden. *O. variegata* wächst oft in individuenreichen Gruppen zusammen.

• WIRT
Schmarotzt vor allem auf *Fabaceae*-Arten.

• GESAMTVERBREITUNG
Vor allem im südwestlichen Mittelmeergebiet; Süd- und Mittelspanien, Südostfrankreich, den südlichen Teil der Schweiz und Italien (auch auf Korsika, Sardinien und Sizilien). Nicht auf den Balearen. Auch in Nordafrika (Algerien). Das Verbreitungsgebiet für Spanien wurde auf Grund vorläufiger Angaben bestimmt.
Die Art is sehr selten, zumindest an ihren nördlichen Standorten.

• BEMERKUNGEN
Im Bearbeitungsgebiet dieses Buches wurde die Art bis jetzt nur in der Schweiz (Kanton Tessin) und in Norditalien gefunden.
Orobanche variegata ist leicht mit *O. gracilis* zu verwechseln. Ein wichtiges Unterscheidungsmerkmal von *O. variegata* ist der Mittellappen der Unterlippe der Blütenkrone, der meistens doppelt so groß ist wie die beiden Seitenlappen.

Variegated Broomrape

• SPECIES DESCRIPTION
The plant is usually stout, approximately 30-70 cm tall. The stem is stout, erect, yellow, yellowish-brown, orange-yellow, reddish, golden yellow, rarely purple, richly (rarely sparsely) glandular-pubescent (sometimes glabrous), densely scaled (imbricate) below, laxly scaled above. The lower scale leaves are broadly lanceolate to oval and glabrous, the upper ones lanceolate and glandular-pubescent, erect to spreading. The inflorescence is cylindrical, with numerous flowers in a dense spike, often lax below; flowers distributed over at least two thirds of the stem. Bracteoles are absent. The bract is about as long as or slightly longer than the corolla, lanceolate and deflexed, yellowish-brown and almost glabrous. The calyx-segments are more or less unequally bidentate, rarely entire, about half to two thirds as long as the corolla, narrowly lanceolate, glandular-pubescent (especially at the base), with tips often more darkly coloured than the corolla. The flowers are of medium size, erect at first to horizontal later. The corolla is 14-23 mm long, campanulate and distinctly inflated above the insertion of the stamens; the outside light to dark yellow (reddish or red-brown at the tips), with short glandular hairs; the inside dark red and slightly veined, glabrous. The dorsal line of the corolla is evenly curved (usually more strongly curved in the middle) and almost straight near the upper lip. The upper lip of the corolla is usually entire or bilobate with wide, porrect, rounded lobes. The lower lip of the corolla is deflexed with three rounded, unevenly crenate lobes, the middle lobe about twice the size of the side lobes. The stamens are inserted 2-4 mm above the base of the corolla-tube. The filaments are more or less pubescent (or glabrous) below and richly to sparsely glandular-pubescent above, up to the anthers. The anthers are pubescent at the line of fusion. The style is glabrous below (ovary) and sparsely glandular-pubescent near the stigma. The stigma consists of two rounded, spherical lobes and is (dark) yellow below and reddish above. The flowers have an unpleasant smell. $2n = 38$.

• FLOWERING TIME
June to July, as early as mid-April in the Mediterranean region.

• HABITAT
Orobanche variegata grows mainly in arid, semi-arid and xerothermic grassland, open thickets (along the edge of forests) and on stony-rocky calcareous slopes in warm, sunny places, on alkaline, loamy and calcareous soil. *O. variegata* often grows in large groups.

• HOST
Parasitic mainly on *Fabaceae* species.

• DISTRIBUTION
Mainly in the south-western Mediterranean region; southern and central Spain, south-eastern France, the southern part of Switzerland and Italy (also on Corsica, Sardinia and Sicily). Not on the Balearic Islands. Also in northern Africa (Algeria). The range on the Spanish distribution map is based on preliminary information.
The species is very rare, at least in its northern locations.

• COMMENTS
Within the area covered by this book *Orobanche variegata* has been found in Switzerland (Canton Ticino) and in northern Italy only.
Orobanche variegata is not easily distinguished from *O. gracilis*. An important distinctive feature is the middle lobe of the lower lip of the corolla, which is usually double the size of the lateral lobes.

OROBANCHE VARIEGATA

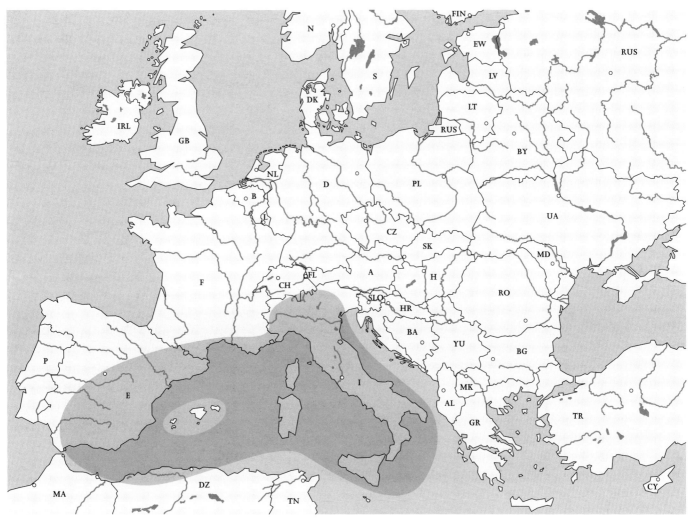

OROBANCHE VARIEGATA

Grazalema, Andalucia (E), 12-4-1994

OROBANCHE VARIEGATA

Grazalema, Andalucia (E), 12-4-1994

Grazalema, Andalucia (E), 12-4-1994

Grazalema, Andalucia (E), 12-4-1994

Grazalema, Andalucia (E), 12-4-1994

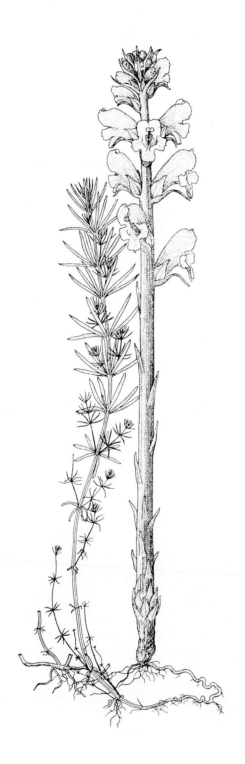 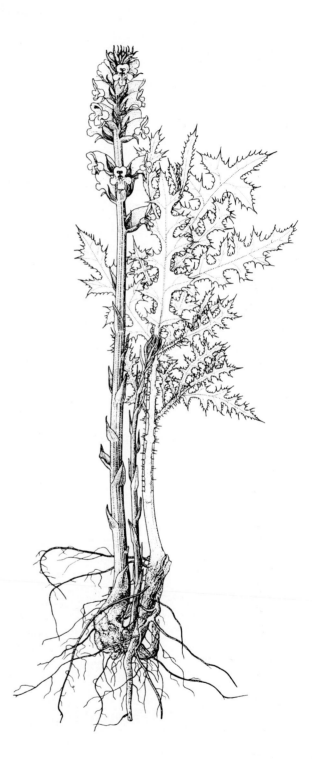

Orobanche caryophyllacea mit Wirtspflanze/with host (*Galium verum*)   *Orobanche pallidiflora* mit Wirtspflanze/with host (*Cirsium vulgare*)

Literatur
Bibliography

ADLER, W., K. OSWALD & R. FISCHER, 1994. Exkursionsflora von Österreich. Verlag Eugen Ulmer; Stuttgart & Wien.

BECK VON MANNAGETTA, G.R., 1882. Die Orobanchen der niederösterreichischen Flora. In: Halácsy & Braun, Nachträge zur Flora von Nieder-Österreich: 115; Wien.

BECK VON MANNAGETTA, G.R., 1890. Monographie der Gattung *Orobanche*. Biblioth. Bot. 19; Cassel.

BECK-MANNAGETTA, G.R., 1927. *Orobanche*. In: Hannig, E. & H. Winkler, Die Pflanzenareale 1 (7): 73-81; Jena.

BECK-MANNAGETTA, G., 1930. *Orobanchaceae* L. In: Engler, A., Das Pflanzenreich IV. (261); Leipzig.

BEDI, J., 1994. Further studies on control of sunflower broomrape with *Fusarium oxysporum* f. sp. *orthroceras* - a potential mycoherbicide. In: Pieterse, A.H., J.A.C. Verkleij & S.J. ter Borg (eds.), Biology and management of *Orobanche*, Proceedings of the Third International Workshop on *Orobanche* and related *Striga* research: 539-544. Royal Tropical Institute; Amsterdam, The Netherlands.

BELDIE, AL., 1967. Flora si vegetatia muntilor Bucegi. Academiei Republicii Socialiste Romania; Bucuresti.

BELDIE, AL., 1979. Flora Romaniei. Vol. II, 120-123. Academiei Republicii Socialiste Romania; Bucuresti.

BENAC, A., 1967. Flora Bosnae et Hercegovinae. Pars 2, 103-109; Sarajevo.

BLAMEY, M. & C. GREY-WILSON, 1993. Mediterranean Wild Flowers. Harper Collins Publishers; Hong Kong.

BOESHORE, I., 1920. The morphological continuity of Scrophulariaceae and Orobanchaceae. Contr. Bot. Lab., Morris Abor. Univ. Pennsylvania 5: 139-177; Philadelphia.

BOLMAN, J., 1967. Bitterkruidbremraap (*Orobanche picridis* Schultz ex Koch). Natura 63: 172-175; Utrecht.

BONAFÈ BARCELÒ, F., 1977-1980. Flora de Mallorca. Volum I: 135-140; Palma de Mallorca.

BORG, S.J. TER (ED.), 1986. Biology and control of *Orobanche*. Proceedings of a workshop in Wageningen, The Netherlands 13-17 January 1986; Wageningen.

BORG, S.J. TER, H. NABER, T.M. BEZEMER & F.M.F. ZAITOUN, 1994. *Orobanche minor* in the Netherlands: An agricultural problem became an endangered species. In: Pieterse, A.H., J.A.C. Verkleij & S.J. ter Borg (eds.), Biology and management of *Orobanche*, Proceedings of the Third International Workshop on *Orobanche* and related *Striga* research: 614-618. Royal Tropical Institute; Amsterdam, The Netherlands.

BORNMÜLLER, 1904. Beitrag zur Kenntnis der Orobanchenflora Vorderasiens. Bull. Herb. Boissier 2. sér. IV, 673; Geneva.

BRANDZA, D., 1879-1883. Prodromul Florei Romane, 161-163; Bucuresti.

CADEVALL DIARS, J. I, 1932. Flora de Catalunya. *Orobanche*, 294-311; Barcelona.

CAMARDA, I., 1983. Proposta di una scheda dei caratteri morfologici del genere "*Orobanche*" L. Webbia 37 (1): 171-184; Firenze.

CELAKOVSKY, L. 1891. Flora von Österreich-Ungarn. I. Böhmen. Österr. Bot. Z. 41: 69-73; Wien.

CHATER, A.O. & D.A. WEBB, 1972. *Orobanche* L. In: Tutin, T.G., V.H. Heywood, N.A. Burges, D.M. Moore, D.H. Valentine, S.M. Walters & D.A. Webb, Flora Europaea 3: 286-293; Cambridge.

CLAMPHAM, A.R., T.G. TUTIN & E.F. WARBURG, 1962. Flora of the British Isles. University Press; Cambridge.

COOLS, J.M.A., 1989. Atlas van de Noordbrabantse Flora. Stichting Uitgeverij KNNV 51; Utrecht.

COSTE, H., 1937. Flore descriptive et illustrée de la France, de la Corse et des contrées limitrophes; Paris.

COSTE, H., 1977. Flore descriptive et illustrée de la France, de la Corse et des contrées limitrophes, Quatrième Supplément; Paris.

COUTINHO, A.X.P., 1939. Flora de Portugal (Plantas vasculares). Bertrand Ltd.; Lisboa.

CVELEV, N.N., 1981. Flora evropejskoj casti SSSR (Flora of the European part of the USSR), Volume V: 319-335; Leningrad.

DAVIS, P.H., 1982. Flora of Turkey. *Orobanchaceae* (90): 1-23, University Press; Edinburgh.

DEGEN, A. VON, 1936-1938. Flora Velebitica; Budapest.

DEMUTH, S., 1992. Über einige seltene *Orobanche*-Arten (*Orobanchaceae*) in Baden-Württemberg. Carolinea 50: 57-66; Karlsruhe.

DIJKSTRA, S.J., 1967. Nog eens de bremraap. Natura 64 (10): 197; Utrecht.

DIJKSTRA, S.J., 1969. Voedselopname bij merkwaardige planten, VIII *Orobanche*. Natuurhist. Maandbl. 58 (12): 190-195; Maastricht.

DÖRR, E., 1978. Flora des Allgäus 12 (*Scrophulariaceae-Cucurbitaceae*). Ber. Bayer. Bot. Ges. 49: 203-270; München.

DOSTAL, J., 1989. Nová Kvetena CSSR. Vol. 2: 950-958, Academia; Praha.

EHRENDORFER, F. (HRSG.), 1973. Liste der Gefäßpflanzen Mitteleuropas, bearbeitet von W. Gutermann, 2. Aufl.; Stuttgart.

EILART, J., M. KASK, V. KUUSK, L. LAASIMER, E. LELLEP, V. PUUSEPP, S. TALLTS & JA. L. VILJASOO, 1973. Eesti NSV Floora V. Zooloogia ja Botaanika Instituut. Valgus; Tallinn.

ENCHEVA, V. & P. SHINDROVA, 1994. Broomrape (*Orobanche cumana* Wallr.) - a hindrance to sunflower production in Bulgaria. In: Pieterse, A.H., J.A.C. Verkleij & S.J. ter Borg (eds.), Biology and management of *Orobanche*, Proceedings of the Third International Workshop on *Orobanche* and related *Striga* research: 619-622. Royal Tropical Institute; Amsterdam, The Netherlands.

FATARE, I., 1992. Vides Aizsardziba Latvija 3: 130, 150, 223; Riga.

FEDOROV, A., 1981. Flora Partis Europaeae URSS V.; Leningrad.

FEINBRUNN-DOTHAN, N., 1978. Flora Palaestina. Part 3: 208-215, The Israel Academy of Sciences and Humanities; Jerusalem.

FITTER, A., 1978. An Atlas of the Wild Flowers of Britain and northern Europe. Collins; London.

FOLEY, M.J.Y., 1992. Some British *Orobanche* variants. B.S.B.I. News 60: 64; London.

FOLEY, M.J.Y., 1993. *Orobanche reticulata* Wallr. populations in Yorkshire (north-east England). Watsonia 19: 247-257; Arbroath.

FORMANEK, E., 1886. Beitrag zur Flora des mittleren und südlichen Mährens; Prag.

FOURNIER, P., 1961. Les quatre Flores de la France. Editions Paul Lechevalier; Paris.

FRANZ, W.R., 1977. Die Violette Sommerwurz, *Orobanche purpurea* JACQ.- eine äußerst seltene Schmarotzerpflanze der Kärnter Flora. Carinthia II: 167/87: 327-332; Klagenfurt.

FREDE, A., 1988. *Orobanche elatior* Sutt. bei Basdorf und Hesperinghausen im Landkreis Waldeck-Frankenberg. Hess. Florist. Briefe 37 (3): 43-44; Darmstadt.

FRITSCH, K., 1922. Exkursionsflora für Österreich und die ehemals österreichischen Nachbargebiete; Wien & Leipzig.

FUCHS-ECKERT, H.P., 1987. Zur Situation von *Orobanche ramosa* Linnaeus (Hanfwürger, Tabaktod, ästige Sommerwurz) in der Schweiz. Jber. Natf. Ges. Graubünden 104: 127-157; Chur.

GILLI, A., 1966. Bestimmungsschlüssel der mitteleuropäischen Varietäten und Formen von *Orobanche*. Verh. Zool.- Bot. Ges.: 105/106: 171-181; Wien.

GILLI, A., 1966. *Orobanchaceae*. In: Hegi, G., Illustrierte Flora von Mitteleuropa 6 (1): 470-505; München.

GILLI, A., 1971/1972. Bemerkenswerte *Orobanche*-Funde aus Niederösterreich. Verh. Zool. Bot. Ges.: 110/111: 5-6; Wien.

GRISEBACH, A., 1844. Florae rumelicae et bithynicae. Volumen secundum: 55-60; Brunsvigae.

GUIMARAES, J.A., 1904. Monografia das Orobanchaceas. Brotéria III: 1-188; Lisboa.

GUINOCHET, M. & R. DE VILMORIN, 1975. Flora de France. Editions du centre de la recherche scientifique 15, quai Anatole-France; Paris.

HAEUPLER, H. & P. SCHÖNFELDER, 1988. Atlas der Farn- und Blütenpflanzen der Bundesrepublik Deutschland. Verlag Eugen Ulmer; Stuttgart.

HALACSY, E. VON, 1902. Conspectus Florae graecae. Volumen II: 444-460; Leipzig.

HAPPEL, E., 1984. Ein Fund der Nelken-Sommerwurz, *Orobanche caryophyllacea* Sm., im westlichen Hohen Vogelsberg. Hess. Florist. Briefe 33 (1): 10-11; Darmstadt.

HAPPEL, E., 1984. Zu dem Beitrag "Ein Fund der Nelken-Sommerwurz, *Orobanche caryophyllacea* Sm., im westlichen Hohen Vogelsberg". Hess. Florist. Briefe 33 (4): 62; Darmstadt.

HARTL, H., G. KNIELY, G.H. LEUTE, H. NIKLFELD & M. PERKO, 1992. Verbreitungsatlas der Farn- und Blütenpflanzen Kärntens. Naturwissenschaftliche Verein für Kärnten; Klagenfurt.

HARTMANN, J., 1988. Gefährliche Schönheiten, Parasitäre Pflanzen- bei uns geschützt, doch in den Entwicklungsländern eine Plage. Die Zeit. K.G. Zeitverlag G. Bucerius; Hamburg.

HAYEK, A., 1914. *Orobanchaceae*. In: Hegi, G., Illustrierte Flora von Mitteleuropa 6 (1): 132-155; München.

HAYEK, A., 1924-1933. Prodromus Florae peninsulae Balcanicae. Band 2, 212-227; Berlin-Dahlem.

HESS, H.E., E. LANDOLT & R. HIRZEL, 1972. Flora der Schweiz und angrenzender Gebiete. Band 3: 250-260, Birkhäuser-Verlag; Basel & Stuttgart.

HIEMEYER, F., 1989. Die Sommerwurzarten (*Orobanche*) in Bayerisch-Schwaben. Ber. Naturwiss. Ver. Schwaben e.V. 93 (2): 27-35; Augsburg.

HOFFMANN, A., 1988. Die Violette Sommerwurz, *Orobanche purpurea* Jacq., im Gladenbacher Bergland. Hess. Florist. Briefe 37 (2): 20-22; Darmstadt.

HOFFMANN, A. & W. KLEIN, 1988. Ein weiterer Fund der Nelken-Sommerwurz, *Orobanche caryophyllacea* Smith, im Gladenbacher Bergland. Hess. Florist. Briefe 37 (3): 42; Darmstadt.

HOHENESTER, A. & W. WELSS, 1993. Exkursionsflora für die Kanarischen Inseln. Verlag Eugen Ulmer; Stuttgart.

HOLMBOE, J., 1914. Studies on the vegetation of Cyprus. Bergens Museums Skrifter. Ny Raekke. Bind 1 (2): 167-169; Bergen.

HULTEN, E., 1950. Atlas of the distribution of vascular plants in NW. Europe. AB Kartografiska Institutet; Stockholm.

HULTEN, E. & H. ANTHON. Snyltrotsfamiljen, *Orobanchaceae*. In: Var Svenska Flora i färg 11: 542. AB Svensk Litteratur; Stockholm.

HULTEN, E. & M. FRIES, 1986. Atlas of north European vascular plants. North of the tropic of cancer. Volume II: 854-857, Koeltz Scientific Books; Königstein.

HULTEN, E. & M. FRIES, 1986. Atlas of north European vascular plants. North of the tropic of cancer. Volume III: 1127, Koeltz Scientific Books; Königstein.

JANCHEN, E., 1977. Flora von Wien, Niederösterreich und Nordburgenland. Verein für Landeskunde von Niederösterreich und Wien; Wien.

JANKEVICIENE, R., 1976. Dzioveklé - *Orobanche* L. In: Lietuvos TSR Flora V: 492-497, 578; Vilnius.

JAUDZEME, V., 1959. Brünkatu Dzimta - *Orobanchaceae* Lindl. In: Galenieka, P., Latvijas PSR Flora IV: 253-258, 493; Riga.

JAVORKA, S. & V. CSAPODY, 1979. Ikonographie der Flora des südöstlichen Mitteleuropa. Gustav Fischer Verlag; Stuttgart.

JONES, M., 1989. Taxonomic and ecological studies in the genus *Orobanche* L. in the British Isles. Ph. D. thesis, University of Liverpool; Liverpool.

JONES, M., 1991. Studies into the Pollination of *Orobanche* species in the British Isles. In: Progress in *Orobanche* Research: 6-17; Tübingen.

JOSIFOVIC, M., 1974. Flora de la Republique Socialiste de Serbie VI: 284-308; Beograd.

KAISER, E., 1950. Die Steppenheiden des mainfränkischen Wellenkalkes zwischen Würzburg und dem Spessart. Ber. Bayer. Bot. Ges. 28: 125-180; München.

KNOCHE, H., 1921-1923. Flora balearica; Montpellier.

KNOTTERS, C., 1994. Hot-spot Distelbremraap. Natura 8: 175-175; Utrecht.

KOCH, L., 1887. Die Entwicklungsgeschichte der *Orobanchacea* mit besonderer Berücksichtigung ihrer Beziehungen zu den Kulturpflanzen; Heidelberg.

KORNECK, D., 1972. *Orobanche elatior* Sutt. in Rheinhessen und Nachbargebieten. Hess. Florist. Briefe 21 (2): 18-20; Darmstadt.

KREEFTENBERG, H., 1990. De Bitterkruidbremraap bij Tolkamer. Natuur en Landschap in Achterhoek en Liemers 4 (3): 90-95. Staring Instituut; Doetinchem.

KREEFTENBERG, H., 1990. De Grote bremraap en de Klavervreter in de omgeving van Doetinchem. Natuur en Landschap in Achterhoek en Liemers 4 (4): 107-110. Staring Instituut; Doetinchem.

KREEFTENBERG, H.G., 1992. Bremrapen in de Liemers. Natura 88 (6): 130-133; Utrecht.

KREUTZ, C.A.J., 1988. *Orobanche elatior* op de Kunderberg. Natura 84 (8): 222-223; Utrecht.

KREUTZ, C.A.J., 1989. De Centauriebremraap (*Orobanche elatior* Sutton) in Zuid-Limburg: een nieuwe soort van de Nederlandse flora. Gorteria 15 (2): 30-31; Leiden.

KUNZ, R., 1957. Orobanchen an der unteren Bergstraße. Hess. Florist. Briefe 71 (1): 1-3; Offenbach.

LAMBINON, J., J.E. DE LANGHE, L. DELVOSALLE & J. DUVIGNEAUD, 1992. Nouvelle Flore de la Belgique, du Grand-Duché de Luxembourg, du Nord de la France et des Régions voisines. Editions du Patrimoine du Jardin botanique national de Belgique; Meise.

LAND, J. VAN DER, 1966. *Orobanchaceae*. In: Ooststroom S.J. van, R. van der Veen, S.E. de Jongh, F.A. Stafleu & V. Westhoff (red.), Flora Neerlandica, Flora van Nederland, deel IV (2), 187-205, KNBV; Amsterdam.

LANGHE, J.E. DE, L. DELVOSALLE, J. DUVIGNEAUD, J. LAMBINON & C. VANDEN BERGEN, 1983. Flora van België, het Groothertogdom Luxemburg, Noord-Frankrijk en de aangrenzende gebieden. Nationale Plantentuin van België; Meise.

LEWEJOHANN, K., 1971. Die Gattung *Orobanche* im südlichen Niedersachsen und angrenzenden Gebieten. Göttinger Florist. Rundbr. 5: 6-10; Göttingen.

LID, J., 1963. Norsk og Svensk Flora. Det Norske Samlaget; Oslo.

LINKE, K.-H., J. SAUERBORN & M.C. SAXENA, 1989. *Orobanche* Field Guide. University of Hohenheim; Filderstadt.

LUDWIG, W., 1969. *Orobanche gracilis* bei Mainz und Gießen. Hess. Florist. Briefe 18 (2): 19-20; Darmstadt.

LUDWIG, W., 1991. Über *Orobanche reticulata* Wallr. (incl. *O. pallidiflora* Wimm. & Grab.) in Hessen. Hess. Florist. Briefe 40 (1): 1-3; Darmstadt.

LUDWIG, W., 1991. Zu *Orobanche elatior* Sutton in der Rhön. Hess. Florist. Briefe 40 (2): 31; Darmstadt.

MADALSKI, J., 1967. *Orobanchaceae*. In: Browicz, K. *et al.*, Flora Polska, Tom XI, 25-53. Warszawa; Krakow.

MADALSKI, J., 1973. Atlas Flory Polskiej I Ziem Osciennych (Florae Poloniсае Terrarumque Adiacentium Iconographia). Tom XVII - ZESZYT 1 (33 Tablice), Scrophulariaceae (Pars 4) *Orobanchaceae*; Warszawa, Wroclaw, Krakow.

MEIJDEN, R. VAN DER, 1990. Heukels' Flora van Nederland, 21ste druk. Wolters-Noordhoff; Groningen.

MEIKLE, R.D., 1985. Flora of Cyprus. Volume 2: 1232-1242, Royal Botanic Gardens; Kew.

MENNEMA, J., A.J. QUENÉ-BOTERENBROOD & C.L. PLATE, 1980. Atlas van de Nederlandse flora deel 1. Uitgestorven en zeer zeldzame plantesoorten. Uitgeverij Kosmos; Amsterdam.

MENNEMA, J., A.J. QUENÉ-BOTERENBROOD & C.L. PLATE, 1985. Atlas van de Nederlandse flora deel 2. Zeldzame en vrij zeldzame planten. Uitgeverij Bohn, Scheltema & Holkema; Utrecht.

MEUSEL, H., E. JÄGER, S. RAUSCHERT & E. WEINERT, 1978. Vergleichende Chorologie der Zentraleuropäischen Flora II (Text): 410-412. Veb Gustav Fischer Verlag; Jena.

MEUSEL, H., E. JÄGER, S. RAUSCHERT & E. WEINERT, 1978. Vergleichende Chorologie der Zentraleuropäischen Flora II (Karten): 412-415. Veb Gustav Fischer Verlag; Jena.

MOSSBERG, B., L. STENBERG & S. ERICSSON, 1992. Den Nordiska Floran. Wahlström & Widstrand; Stockholm.

MURBECK, S., 1891. Beiträge zur Kenntnis der Flora von Südbosnien und der Hercegovina; Lund.

MUSSELMANN, L.J., 1986. Taxonomy of *Orobanche*. In: Borg, S.J. ter (ed.), Proceedings of a workshop on biology and control of *Orobanche*. LH/VPO: 2-10; Wageningen.

NIESCHALK, A., 1954. Die Hohe Sommerwurz (*Orobanche major* L.) im westfälisch-waldeckischen Grenzgebiet. Natur & Heimat 14 (1): 25-26; Münster.

NIESCHALK, A., 1968. *Orobanche libanotidis* (= *O. bartlingii* Grisebach) in Hessen. Hess. Florist. Briefe 17 (3): 35-42; Darmstadt.

NIESCHALK, A. & CH. NIESCHALK, 1974. Mitteilungen zur Verbreitung von *Orobanche bartlingii* Grisebach (=*Orobanche libanotidis* Ruprecht, *O. alsatica* Kirschleger var. *libanotidis* [Ruprecht] Beck) in Bayern. Ber. Bayer. Bot. Ges. 45: 71-74; München.

NOVOPOKROVSKIJ, I.V. & N.N. CVELEV, 1950. Zarazichovye - *Orobanchaceae* Lindl. In: Komarov, V.L., 1958. Flora SSSR 23: 19-117; Moskva & Leningrad.

OBERDORFER, E., 1970. Pflanzensoziologische Exkursionsflora für Süddeutschland und die angrenzenden Gebiete. 3. Aufl., Verlag Eugen Ulmer; Stuttgart.

OBERDORFER, E., 1983. Pflanzensoziologische Exkursionsflora. 5. Aufl., Verlag Eugen Ulmer; Stuttgart.

OBERDORFER, E., 1990. Pflanzensoziologische Exkursionsflora. 5. Aufl., Verlag Eugen Ulmer; Stuttgart.

OLIVEIRA-VELLOSO, J.A.R., A. PUJADAS-SALVA & E. HERNÁNDEZ-BERMEJO, 1994. The Genus *Orobanche* in the crops of Andalusia (South of Spain). In: Pieterse, A.H., J.A.C. Verkleij & S.J. ter Borg (eds.), Biology and management of *Orobanche*, Proceedings of the Third International Workshop on *Orobanche* and related *Striga* research: 628-634. Royal Tropical Institute; Amsterdam, The Netherlands.

PERRING, F.H. & S.M. WALTERS, 1962. Atlas of the British flora; London, Edinburgh.

PHILP, E.G., 1982. Atlas of the Kent flora; Maidstone.

PIETERSE, A.H., J.A.C. VERKLEIJ & S.J. TER BORG (EDS.), 1994. Biology and management of *Orobanche*, Proceedings of the Third International Workshop on *Orobanche* and related *Striga* research, 8-12 November, 1993. Royal Tropical Institute; Amsterdam, The Netherlands.

PIGNATTI, S., 1982. Flora d'Italia, *Orobanchaceae*: 606-616; Bologna.

POLUNIN, O., 1980. Flowers of Greece and the Balkans, a field guide. Oxford University Press; Oxford.

POST, G.E., 1933. Flora of Syria, Palestine and Sinai. Volume II: 312-316, America Press; Beirut.

PUGSLEY, H.W., 1926. The British *Orobanche* list. J. Bot. 64: 16-19; London.

PUGSLEY, H.W., 1940. Notes on *Orobanche* L. J. Bot. 78: 105-116; London.

PUJADAS-SALVA, A., E. HERNÁNDEZ-BERMEJO & J.A.R. OLIVEIRA-VELLOSO, 1994. The Genus *Orobanche* in Andalusia (southern Spain): Taxonomical, Chorological & Ecological aspects. In: Pieterse, A.H., J.A.C. Verkleij & S.J. ter Borg (eds.), Biology and management of *Orobanche*, Proceedings of the Third International Workshop on *Orobanche* and related *Striga* research: 132-137. Royal Tropical Institute; Amsterdam, The Netherlands.

PUSCH, J., 1989. Wiederfund des seltenen Quendel-Sommerwurz (*Orobanche alba* Steph. ex Willd.) im Nakken südlich von Bad Frankenhausen. Mitt. Florist. Kartierung 15 (1/2): 82-83; Halle.

PUSCH, J., 1990 (1989). Die Sommerwurzarten des Kreises Artern; Bad Frankenhausen.

PUSCH, J., & K.-J. BARTHEL, 1988. Über aktuelle und ehemalige Vorkommen von Orobanchen im Kyffhäusergebirge. Mitt. Florist. Kartierung 14 (1/2): 13-29; Halle.

PUSCH, J., & K.-J. BARTHEL, 1990. Zum Vorkommen und zur Gefährdung der Sommerwurzarten in Thüringen. Landschaftspflege Naturschutz Thüringen 27 (4): 90-95; Suhl.

PUSCH, J., & K.-J. BARTHEL, 1992. Über Merkmale und Verbreitung der Gattung *Orobanche* L. in den östlichen Bundesländern Deutschlands. Gleditschia 20 (1): 33-56; Berlin.

PUSCH, J., & K. WEIKERT, 1991. Die Belege der Gattung *Orobanche* L. (Sommerwurzgewächse) im Herbarium des Naturkundemuseums Erfurt. Veröff. Naturkundemuseum Erfurt 10: 13-15; Erfurt.

RAABE, U., 1987. Quendel-Sommerwurz, *Orobanche alba* Steph. ex Willd., und Hohe Sommerwurz, *Orobanche elatior* Sutton, im Raum Brilon, Hochsauerlandkreis. Florist. Rundbr. 21 (1): 51-53; Göttingen.

RAABE, U. & R. GÖTTE, 1989. Die Bleiche Distel-Sommerwurz, *Orobanche reticulata* subsp. *pallidiflora*, in Westfalen. Florist. Rundbr. 23 (1): 15-16; Bochum.

RECHINGER, K.H., 1943. Flora aegaea, Flora der Inseln und Halbinseln des Ägäischen Meeres: 487-491. Springer-Verlag; Wien.

REICHENBACH, H.G.L., 1829. Iconographia botanica VII; Leipzig.

REICHENBACH, H.G., 1862. Icones florae germanicae et helveticae XX; Leipzig.

REUTER, G.F., 1847. *Orobanchaceae*. In: De Candolle, Prodromus syst. nat. regni vegetabilis XI; Parisiis.

ROLLÁN, M.G., 1985. Claves de la Flora de Espana (Peninsula y Baleares); Madrid.

ROMPAEY, E. VAN & L. DELVOSALLE, 1978. Atlas van de Belgische en Luxemburgse flora. Tekstgedeelte; Meise.

Rompaey, E. van & L. Delvosalle, 1979. Atlas van de Belgische en Luxemburgse flora 2; Meise.

Rostrup & Jorgensen, 1973. Den danske flora. Gyldendal; Kobenhavn.

Rothmaler, W., R. Schubert, E. Jäger & K. Werner, 1988. Exkursionsflora für die Gebiete der DDR und der BRD. Band 3, Atlas der Gefäßpflanzen. Volk und Wissen Volkseigener Verlag; Berlin.

Rothmaler, W., R. Schubert & W. Vent, 1988. Exkursionsflora für die Gebiete der DDR und der BRD. Band 4, Kritischer Band. Volk und Wissen Volkseigener Verlag; Berlin.

Rout, G.C.C., 1927. Conspectus de la Flore de France; Paris.

Rumsey, F.J. & S.L. Jury, 1991. An account of *Orobanche* L. in Britain and Ireland. Watsonia 18: 257-295; Arbroath.

Saule, M., 1991. La grande Flore illustrée des Pyrénées; Éditions Milan.

Savulescu, T. & E.I. Savulescu, 1963. Nyarady. Flora Reipublicae Popularis Romanicae. Vol. VIII, 33-72; Bucuresti.

Schieferdecker, K., 1939. *Orobanche picridis* F. Schultz auf dem Knebel bei Hildesheim. Hercynia 1: 488-589; Halle (Saale).

Schlesinger, S., 1991. Zweiter Fund von *Orobache picridis* F.W. Schultz in Baden-Württemberg. Carolinea 49: 125; Karlsruhe.

Schmeil-Fitschen, 1967. Flora von Deutschland und seinen angrenzenden Gebieten. 83. Auflage, bearbeitet von W. Rauh & K. Senghas, Quelle & Meyer Verlag; Heidelberg.

Schönfelder, P., A. Bresinsky, E. Garnweidner, E. Krach, H. Linhard, O. Mergenthaler, W. Nezadal & V. Wirth, 1990. Verbreitungsatlas der Farn- und Blütenpflanzen Bayerns. Verlag Eugen Ulmer; Stuttgart.

Schultz, F.W., 1829. Beitrag zur Kenntnis der deutschen Orobanchen; Speyer.

Seithe, A., 1972. *Orobanchaceae*. In: Garke, A., Illustrierte Flora von Deutschland und angrenzende Gebiete. 23. Aufl, Verlag Paul Parey; Berlin & Hamburg.

Shindrova, P., 1994. Distribution and race composotion of *Orobanche cumana* Wallr. in Bulgaria. In: Pieterse, A.H., J.A.C. Verkleij & S.J. ter Borg (eds.), Biology and management of *Orobanche*, Proceedings of the Third International Workshop on *Orobanche* and related *Striga* research: 142-145. Royal Tropical Institute; Amsterdam, The Netherlands.

Sluschny, H., G. Matthes & J. Duty, 1985. Die Bitterkraut-Sommerwurz (*Orobanche picridis* F.W. Schultz) neu in Mecklenburg. Bot. Rundbrief Bez. Neubrandenburg 17: 53-58; Neustrelitz.

Soó, R., 1968. Magyar flora és vegetacio rendszertani-növényföldrajzi kézikönyve. Vol III, 237-250. Akedémiai kiado; Budapest.

Stace, C., 1975. New Flora of the British Isles. Cambridge University Press; Cambridge.

Sterner, R., 1924. Om Ölands flora I, *Orobanche purpurea* Jacq. funnen pa Öland. Svensk. Bot. Tidskr. 18, 465; Lund.

Stojanow, I., B. Stefanow & B. Kitanov, 1966-1967. Flora na Balgarija; Sofija.

Straka, K., 1994. Ein neuer Wuchsort von *Orobanche arenaria* Borkh. im Stadtgebiet von Mainz. Hess. Florist. Briefe 43 (2): 32; Darmstadt.

Strid, A. & K. Tan, 1991. Mountain flora of Greece. Volume two, 263-276. University Press; Edinburgh.

Süssenguth, K. & K. Ronniger, 1942. Über *Orobanche alsatica* Kirschl. var. *Mayeri* Ssg. et Ronniger. Eine neue *Orobanche* aus der Schwäbischen Alb. Beitr. Naturk. Forsch. Oberrheingebiet 7: 123-127; Karlsruhe.

Swart, J., 1972. Een bijzondere plant op een bijzondere plaats. Natura 68: 63-64; Utrecht.

Uhlich, H., 1990. Zur Verbreitung der Gattung *Orobanche* L. in Sachsen. Sächs. Florist. Mitt. 1: 30-43.

Uhlich, H., J. Pusch & J. Zázworka, 1990. *Orobanche picridis* F.W. Schultz ex Koch nach über 50 Jahren am Radobyl (Böhmisches Mittelgebirge) wiederentdeckt. Severoces. Prir. Litomerice 24: 23-27; Litomerice.

Valdés, B., 1986. *Orobanchaceae*. In: Valdés *et al.* (eds.), Flora Vascular de Andalucia Occidental, Vol. 2: 550-558. Ketrés; Barcelona.

Vaucher, J.P., 1827. Monographie des Orobanches; Genève, Paris.

Velenovsky, J., 1898. Flora Bulgarica. Descriptio et enumeratio systematica plantarum vascularium in principatu bulgariae sponte nascentium; Pragae.

Viljasoo, L., B. Cepurite & Z. Sinkeviciene. In: Flora of the Baltic countries II: 348-351, in Vorbereitung.

Wallroth, F.G., 1825. Orobanches generis diaskeue.

Webb, D.A., 1966. The Flora of European Turkey. Proc. Roy. Irish Acad. 65B: 1-100; Dublin.

Webb, D.A., 1967. An Irish Flora; Dundalk.

Weber, A., 1976. Die Chromosomenzahlen der in Mitteleuropa vorkommenden Arten von *Orobanche* sect. *Orobanche*. Plant Syst. Evol. 124: 303-308, Springer-Verlag; Wien.

Weber, H.-C., 1978. Schmarotzer, Pflanzen die von anderen leben. Belser Verlag; Stuttgart.

Weber, H.-C., 1993. Parasitismus von Blütenpflanzen. Wissenschaftliche Buchgesellschaft; Darmstadt.

Weeda, E.J., 1988. Bremraapfamile/*Orobanchaceae*. In: Weeda, E.J., R. Westra, Ch. Westra & T. Westra, Oecologische Flora, wilde planten en hun relaties, deel 3: 238-245; Deventer.

Wegmann, K. & L.J. Musselman (eds.), 1991. Progress in Orobanche research. Proceedings of the International Workshop on Orobanche research, Obermarchtal, FRG, August 19-22, 1989; Tübingen.

Welten, M. & H.C.R. Sutter, 1982. Verbreitungsatlas der Farn- und Blütenpflanzen der Schweiz. Birkhäuser Verlag, Basel, Boston; Stuttgart.

Wever, A. den, 1918. *Orobanchaceae*. Jaarb. Natuurh. Genootschap in Limburg: 39-44; Maastricht.

Zajac, A. & M. Zajac, 1992. Atlas Rozmieszczenia Roslin Naczyniowych w Polsce (Atpol). Instytutu Botaniki Uniwersytetu Jagiellonskiego; Kraków. Index of general distribution maps vascular plants of Poland. Polish Academy of Sciences. W. Szafer Institute of Botany; Krakow.

Zajac, M., 1992. Index of general distribution maps vascular plants of Poland. Polish Academy of Sciences. W. Szafer Institute of Botany; Kraków.

Zander, O., 1930. Schmarotzende Pflanzen. Brehm Verlag, Band 5; Berlin.

Zázvorka, J., 1984. *Orobanche coerulescens* v Ceskoslovensku. Severoces. Prír. Litomerice 16: 1-23; Litomerice.

Zázvorka, J., 1989. Broomrapes (*Orobanche* L. s.l.) in the Ceské stredohori Mts. Severoces. Prír. Litomerice 23: 19-54; Litomerice.

Zázvorka, J., Orobachaceae. In: Bertova, L. & K. Goliasova., Flora Slovenska, in Vorbereitung; Bratislava.

Ziegenspeck, H., 1957. Zur Verbreitungsbiologie unserer Sommerwurzarten (Orobanchen). Abh. Naturwiss. Vereins Schwaben 12: 27-45; Augsburg.

Zimmermann, A., G. Kniely, H. Melzer, W. Maurer & R. Höllriegl, 1989. Atlas gefährdeter Farn- und Blütenpflanzen der Steiermark. Mitt. Abt. Bot. Landesmus. Joanneum 18/19; Graz.

Inhaltsverzeichnis
Index

6.1 VERZEICHNIS DER SOMMERWURZNAMEN

- 60 Amethyst-Sommerwurz *Orobanche amethystea*
- 44 Ästige Sommerwurz *Orobanche ramosa*
- 68 Bartlings (Heilwurz-) Sommerwurz *Orobanche bartlingii*
- 128 Bitterkraut-Sommerwurz *Orobanche picridis*
- 80 Bläuliche Sommerwurz *Orobanche coerulescens*
- 124 Bleiche Distel- (Netzige) Sommerwurz *Orobanche pallidiflora*
- 148 Bunte Sommerwurz *Orobanche variegata*
- 104 Efeu-Sommerwurz *Orobanche hederae*
- 136 Eigentliche Distel- (Netzige) Sommerwurz *Orobanche reticulata*
- 52 Elsässer (Haarstrang-) Sommerwurz *Orobanche alsatica*
- 144 Gamander-Sommerwurz *Orobanche teucrii*
- 116 Gelbe (Rötlichgelbe) Sommerwurz *Orobanche lutea*
- 84 Gezähnelte Sommerwurz *Orobanche crenata*
- 132 Ginster-Sommerwurz *Orobanche rapum-genistae*
- 92 Große Sommerwurz *Orobanche elatior*
- 112 Hain- (Berberitzen-) Sommerwurz *Orobanche lucorum*
- 96 Hellgelbe (Pestwurz-) Sommerwurz *Orobanche flava*
- 120 Kleine Sommerwurz *Orobanche minor*
- 72 Labkraut- (Gemeine) Sommerwurz *Orobanche caryophyllacea*
- 108 Laserkraut- (Bergkummel-) Sommerwurz *Orobanche laserpitii-sileris*
- 56 Mayer's Sommerwurz *Orobanche alsatica* subsp. *mayeri*
- 76 Nickende Sommerwurz *Orobanche cernua*
- 64 Panzer- (Beifuß-) Sommerwurz *Orobanche artemisiae-campestris*
- 40 Purpur- (Violette) Sommerwurz *Orobanche purpurea*
- 140 Salbei-Sommerwurz *Orobanche salviae*
- 32 Sand-Sommerwurz *Orobanche arenaria*
- 88 Sonnenblumen-Sommerwurz *Orobanche cumana*
- 48 Weiße (Quendel-) Sommerwurz *Orobanche alba*
- 36 Weißwollige (Blaugraue) Sommerwurz *Orobanche caesia*
- 100 Zierliche (Blutrote) Sommerwurz *Orobanche gracilis*

6.2 INDEX OF BROOMRAPE NAMES

- 52 Alsatian Broomrape *Orobanche alsatica*
- 60 Amethyst Broomrape *Orobanche amethystea*
- 112 Barberry Broomrape *Orobanche lucorum*
- 68 Bartling's Broomrape *Orobanche bartlingii*
- 72 Bedstraw (Clove-scented) Broomrape *Orobanche caryophyllacea*
- 36 Blue-grey Broomrape *Orobanche caesia*
- 80 Bluish Broomrape *Orobanche coerulescens*
- 44 Branched (Hemp) Broomrape *Orobanche ramosa*
- 96 Butterbur Broomrape *Orobanche flava*
- 84 Carnation-scented (Bean) Broomrape *Orobanche crenata*
- 120 Common (Small, Lesser) Broomrape *Orobanche minor*
- 144 Germander Broomrape *Orobanche teucrii*
- 132 Greater Broomrape *Orobanche rapum-genistae*
- 104 Ivy Broomrape *Orobanche hederae*
- 108 Laserpitium Broomrape *Orobanche laserpitii-sileris*
- 56 Mayer's Broomrape *Orobanche alsatica* subsp. *mayeri*
- 76 Nodding Broomrape *Orobanche cernua*
- 64 Oxtongue Broomrape *Orobanche artemisiae-campestris*
- 124 Pale Thistle Broomrape *Orobanche pallidiflora*
- 128 Picris Broomrape *Orobanche picridis*
- 140 Sage Broomrape *Orobanche salviae*
- 32 Sand Broomrape *Orobanche arenaria*
- 100 Slender Broomrape *Orobanche gracilis*
- 88 Sunflower Broomrape *Orobanche cumana*
- 92 Tall (Knapweed) Broomrape *Orobanche elatior*
- 136 Thistle Broomrape *Orobanche reticulata*
- 48 Thyme (Red) Broomrape *Orobanche alba*
- 148 Variegated Broomrape *Orobanche variegata*
- 40 Yarrow (Purple) Broomrape *Orobanche purpurea*
- 116 Yellow Broomrape *Orobanche lutea*

6.3 INDEX VAN BREMRAAPNAMEN

- 68 Bartlings bremraap *Orobanche bartlingii*
- 128 Bitterkruidbremraap *Orobanche picridis*
- 80 Blauwachtige bremraap *Orobanche coerulescens*
- 40 Blauwe bremraap *Orobanche purpurea*
- 96 Bleekgele bremraap *Orobanche flava*
- 124 Bleke Distelbremraap *Orobanche pallidiflora*
- 100 Bloedrode bremraap *Orobanche gracilis*
- 148 Bonte bremraap *Orobanche variegata*
- 92 Centauriebremraap *Orobanche elatior*
- 136 Echte Distelbremraap *Orobanche reticulata*
- 52 Elzasser bremraap *Orobanche alsatica*
- 144 Gamanderbremraap *Orobanche teucrii*
- 84 Gekartelde bremraap *Orobanche crenata*
- 64 Gepantserde bremraap *Orobanche artemisiae-campestris*
- 132 Grote bremraap *Orobanche rapum-genistae*
- 44 Hennepvreter *Orobanche ramosa*
- 120 Klavervreter *Orobanche minor*
- 104 Klimopbremraap *Orobanche hederae*
- 76 Knikkende bremraap *Orobanche cernua*
- 108 Laserpitium-bremraap *Orobanche laserpitii-sileris*
- 56 Mayers bremraap *Orobanche alsatica* subsp. *mayeri*
- 116 Rode bremraap *Orobanche lutea*
- 140 Saliebremraap *Orobanche salviae*
- 48 Tijmbremraap *Orobanche alba*
- 36 Viltige bremraap *Orobanche caesia*
- 60 Violette bremraap *Orobanche amethystea*
- 72 Walstrobremraap *Orobanche caryophyllacea*
- 32 Zandbremraap *Orobanche arenaria*
- 88 Zonnebloembremraap *Orobanche cumana*
- 112 Zuurbesbremraap *Orobanche lucorum*

6.4 INDEX OF SCIENTIFIC NAMES

- 48 *Orobanche alba*
- 52 *Orobanche alsatica*
- 56 *Orobanche alsatica* subsp. *mayeri*
- 60 *Orobanche amethystea*
- 32 *Orobanche arenaria*
- 64 *Orobanche artemisiae-campestris*
- 68 *Orobanche bartlingii*
- 36 *Orobanche caesia*
- 72 *Orobanche caryophyllacea*
- 76 *Orobanche cernua*
- 80 *Orobanche coerulescens*
- 84 *Orobanche crenata*
- 88 *Orobanche cumana*
- 92 *Orobanche elatior*
- 96 *Orobanche flava*
- 100 *Orobanche gracilis*
- 104 *Orobanche hederae*
- 108 *Orobanche laserpitii-sileris*
- 112 *Orobanche lucorum*
- 116 *Orobanche lutea*
- 120 *Orobanche minor*
- 124 *Orobanche pallidiflora*
- 128 *Orobanche picridis*
- 40 *Orobanche purpurea*
- 44 *Orobanche ramosa*
- 132 *Orobanche rapum-genistae*
- 136 *Orobanche reticulata*
- 140 *Orobanche salviae*
- 144 *Orobanche teucrii*
- 148 *Orobanche variegata*

Orobanche minor, Prizren, Kosovo (YU), 12-7-1987

Orobanche minor, Landgraaf, Zuid-Limburg (NL), 18-7-1993

Orobanche minor, Casares, Sierra Bermeja (E), 14-4-1994

Orobanche minor, St. Florent, Corse (F), 19-4-1990

Danksagung

Herzlich danken möchte ich:

Herrn J. Hermans (Linne, Niederlande) für das Anfertigen von einigen Zeichnungen,
Herrn J. Klerkx (Maastricht, Niederlande) für die Korrekturlesung in Englisch,
Herrn K. Morschek (Moers, Deutschland) für die Korrekturlesung in Deutsch,
Dr. M. Jones (Newman College, Birmingham, England), Herrn J. Cortenraad (Maastricht, Niederlande) und Herrn A. Hemp (Bayreuth, Deutschland) für die kritische Durchsicht des Manuskriptes,
Herrn J. Zázvorka (Pruhonice, Tschechische Republik) für die Verbreitungskarten der Orobanchen der Tschechischen Republik und der Slowakei, Frau V. Kuusk (Tartu, Estland) für Verbreitungskarten von Estland und Herrn A. Zajac (Kraków, Polen) für diese von Polen, die bisher noch nicht publiziert wurden,
Herrn W.J.P. Meijs für Planung und Marketing,
Prof. Dr. L. Stuchlik vom W. Szafer Institute of Botany (Polish Academy of Sciences, Krakow, Polen) für die Genehmigung zur Übernahme einiger Zeichnungen aus Madalski, J., 1973. *Atlas Flory Polskiej I Ziem Osciennych (Florae Polonicae Terrarumque Adiacentium Iconographia)*, Tom XVII - ZESZYT 1 (33 Tablice), *Scrophulariaceae* (Pars 4) *Orobanchaceae*, Warszawa - Wroclaw - Krakow und dem Blackwell Wissenschafts-Verlag GmbH (Berlin, Deutschland) für die Genehmigung zur Übernahme des Bestimmungsschlüssels aus Hegi, G., *Illustrierte Flora von Mitteleuropa*, 6, 475-477, 1966.

Weiterhin möchte ich nicht versäumen, allen Personen und Institutionen, die in irgendeiner Form am Gelingen des Buches beteiligt waren, für ihre Unterstützung zu danken.

Acknowledgements

The author would like to express his gratitude to:

Mr. J. Hermans (Linne, The Netherlands) who has prepared some of the drawings;
Mr. J. Klerkx (Maastricht, The Netherlands) for editing the English text;
Mr. K. Morschek (Moers, Germany) for checking the German text;
Dr. M. Jones (Newman College, Birmingham, England), Mr. J. Cortenraad (Maastricht, The Netherlands) and Mr. A. Hemp (Bayreuth, Germany) for critically reading the manuscript;
Mr. J. Zázvorka (Pruhonice, Czech Republic) for providing the *Orobanche* distribution maps for the Czech Republic and Slovakia, Mrs. V. Kuusk (Tartu, Estonia) for providing the *Orobanche* distribution maps for Estonia and Mr. A. Zajac (Kraków, Poland) for previously unpublished Polish distribution maps;
Mr. W.J.P. Meijs for his help in planning and marketing the book;
Prof. Dr. L. Stuchlik of the W. Szafer Institute of Botany (Polish Academy of Sciences, Krakow, Poland) for authorizing the reproduction of some drawings from Madalski, J., 1973. *Atlas Flory Polskiej I Ziem Osciennych (Florae Polonicae Terrarumque Adiacentium Iconographia)*, Tom XVII - ZESZYT 1 (33 Tablice), *Scrophulariaceae* (Pars 4) *Orobanchaceae* (Warszawa - Wroclaw - Krakow) and The Blackwell Wissenschafts-Verlag GmbH (Berlin, Germany) for authorizing the reproduction of the identification key from Hegi, G., *Illustrierte Flora von Mitteleuropa*, 6, 475-477 (1966).

In addition, a debt of gratitude is due to all persons and institutions who have contributed in any way to this book.

KOLOPHON/COLOFON

AUTOR/AUTHOR	C.A.J. Kreutz
FOTOGRAPHIE/FOTOGRAPY	C.A.J. Kreutz
SCHLUSSREDACTION/FINAL EDITING	Drs. J. van der Coelen, Drs. B.G. Graatsma
GRAFISCHE GESTALTUNG/DESIGN	Stefan Graatsma, Maastricht
PRODUKTION/PRODUCTION	Bureau van de Manakker bv, Maastricht
LITHOGRAPHIE/LITHOGRAPHY	Micro Color Service, Roermond
DRUCKBEGLEITUNG/PRINT SUPPORT	Studio Beeldwerk, Haarlem
DRUCK/PRINTED BY	De Boer Cuperus, Utrecht
BINDUNG/BOUND BY	Callenbach, Nijkerk
GEDRUCKT AUF/PRINTED ON	Royal Impression Brilliant (135 grs) KNP LEYKAM te Maastricht, NL
AUSGABE/EDITION	Stichting Natuurpublicaties Limburg, postbus 882, 6200 AW Maastricht, NL
	Natuurhistorisch Genootschap in Limburg
ISBN	90-74508-05-7

Februar 1995/february 1995